Lecture Notes in Biomathematics

Managing Editor: S. Levin

28

Paul C. Fife

Mathematical Aspects of Reacting and Diffusing Systems

Springer-Verlag
Berlin Heidelberg New York 1979

Author
Paul C. Fife
Mathematics Department
University of Arizona
Tucson, Arizona 85721/U.S.A.

AMS Subject Classifications (1970): 35B05, 35B40, 35K55, 92A05, 92A15

ISBN 3-540-09117-3 Springer-Verlag Berlin Heidelberg New York
ISBN 0-387-09117-3 Springer-Verlag New York Heidelberg Berlin

Printing and binding: Beltz Offsetdruck, Hemsbach/Bergstr.
2141/3140-543210

pg 11, 12, 23, 37, 58

Sec 2.6

TABLE OF CONTENTS

PREFACE AND GENERAL INTRODUCTION

Modeling and analyzing the dynamics of chemical mixtures by means of differential equations is one of the prime concerns of chemical engineering theorists. These equations often take the form of systems of nonlinear parabolic partial differential equations, or reaction-diffusion equations, when there is diffusion of chemical substances involved. A good overview of this endeavor can be had by reading the two volumes by R. Aris (1975), who himself was one of the main contributors to the theory. Enthusiasm for the models developed has been shared by parts of the mathematical community, and these models have, in fact, provided motivation for some beautiful mathematical results.

There are analogies between chemical reactors and certain biological systems. One such analogy is rather obvious: a single living organism is a dynamic structure built of molecules and ions, many of which react and diffuse. Other analogies are less obvious; for example, the electric potential of a membrane can diffuse like a chemical, and of course can interact with real chemical species (ions) which are transported through the membrane. These facts gave rise to Hodgkin's and Huxley's celebrated model for the propagation of nerve signals. On the level of populations, individuals interact and move about, and so it is not surprising that here, again, the simplest continuous space-time interaction-migration models have the same general appearance as those for diffusing and reacting chemical systems.

The principal ingredients of these models are equations of the form

$$\frac{\partial u}{\partial t} = D\Delta u + F(u),$$

where u is an n-vector (each component represents a measure of one of the diffusing quantities), D is a matrix, and Δ is the Laplace operator in the spatial coordinates. The vector F is a catch-all term describing all reactions and interactions. These systems (and obvious generalizations of them made by allowing for space-time dependence in F and D, nonisotropicity of the diffusion process, and/or "drift" terms) are our object of study. The purpose here is to describe fundamental parts of the theory of these systems. Since they are a tool common to many disciplines, the results given here should be reflected by phenomena from different areas. Nevertheless, it will be clear that the examples given, comments made, and topics chosen, are slanted toward possible biological applications. This reflects the general interest which has developed lately in using reaction-diffusion equations to gain some insight into the dynamics and structure of biological communities and organisms. A good example of this interest is seen in the monograph by Nicolis and Prigogine (1976), which provides a good overview of the study of self-organizing phenomena in many sciences, including biology. It is clear that reaction-diffusion equations are very relevant here. Self-organizing phenomena gave rise (Prigogine and Nicolis 1967; Glansdorff and Prigogine 1971) to the idea of "dissipative structure," which in reaction-diffusion contexts can be modeled by stable solutions of the equations, which are not constant in space-time. In the present notes, Chapters 6, 7, 8, and part of 4 are principally concerned with such stable nonuniform solutions.

The first two chapters take up the question of how reaction-diffusion systems arise as models (Chapter 2 is a case study of such models in population genetics). The remainder of the book is on mathematical methods for handling the equations. Chapter 3 gives some preliminaries. A reasonably complete treatment is given, in Chapter 4, of a single nonlinear diffusion equation in one space dimension, when there are two stable rest states (this is the only case when such an equation supports stable solutions which do not evolve to constant states). Chapters 5-8 are concerned with systems of more than one equation, and describe many of the methods (with some important gaps) now available for their analysis.

Throughout the text (including Chapter 9, where work in some areas not covered by the book is mentioned), references are given to literature on reaction-diffusion systems. These references, again, have gaps, and the most notable ones are the sparse mention of (1) the vast chemical engineering literature, and (2) the mathematical literature dealing with boundary value problems for nonlinear second order ordinary or elliptic differential equations (these give steady-state solutions of scalar reaction-diffusion equations). It would have been impossible to do justice to these areas in this volume.

There exist survey articles and books which, at least in part, touch upon the topics discussed here. The books by Aris and by Nicolis and Prigogine were mentioned above. Diekmann and Temme (1976) provides good expositions of wave fronts for scalar equations, Lyapunov methods for stability, the "Brusselator" model, and the equations modeling signal propagation on a nerve axon. Most of these same subjects, together with others, such as various chemical, porous medium, and epidemic problems, are very well covered in Aronson (1976). The book by Tyson (1976) provides background on the mathematics and chemistry of the best-known laboratory example of a reaction supporting oscillations and signal propagation. Levin (1976a) discusses the phenomenon of patterning in biological communities. Henry (1976) gives a mathematical theory of reaction-diffusion systems from a somewhat alternate point of view. Gavalas (1968) provides background mathematics for the study of chemical systems. Fife (1978b) is a survey article on long-time behavior of RD systems. Murray (1977) contains in-depth discussions of several biological models relevant to the present notes. Mollison (1976) presents a very readable review of models for the spatial spread of epidemics and other population traits. That paper is especially relevant to the present work because of its critical comparison of deterministic and stochastic models.

Notes on which this volume is based were originally written for a course on reaction-diffusion systems at the University of Arizona in Spring, 1977. I have benefited in their preparation from discussions with a number of people, including Warren Ferguson, Thomas Nagylaki, and Simon Rosenblatt. In particular, Thomas Nagylaki offered invaluable advice regarding Chapter 2. I am indebted to Louise Fields for her outstanding performance in typing the manuscript.

Support for these notes was provided by N.S.F. Grant MCS77-02139.

1. MODELING CONSIDERATIONS

The equations of reaction and diffusion are often used to model multispecies populations whose individuals interact in some way to produce other individuals (or eliminate individuals), and also move about in some random manner. The use of these equations is rightly subject to dispute in many cases, and so it is appropriate that we spend some effort in attempting to elucidate the assumptions on which they are based in population dynamical contexts.

Although the language here suggests biological populations, the argument is also intended to cover populations of molecules and ions, in which case "species" means chemical species.

The goal here will be to spell out, and comment upon, a set of general postulates about the population dynamical phenomenon studied, from which equations of reaction-diffusion type may be deduced. Because I wish the arguments to apply to a wide range of population types, reaction mechanisms, and migration rules, I shall say only what is necessary and basic about the specifics of these things. In short, I shall try to make the set of assumptions minimal.

The given assumptions lead rigorously to the RD model, but they do not constitute the only possible avenue to these equations. Colony models (Section 1.9) are sometimes used; see Skellam (1951) for a related random-walk approach. Our postulates are conservative in the sense that the RD systems are also often used to model situations where the assumptions are not fully met, and in many of these cases the model is no doubt still adequate for some purposes.

It should also be brought out that the dynamics of reacting and migrating populations have been successfully modeled without the use of RD systems, and the latter equations are not claimed to constitute the universally best model. For example, the effects one wants to examine may be essentially stochastic, whereas the evolution processes generated by these equations are deterministic. On the other hand, if it is found that the phenomenon studied can be reasonably modeled by equations of reaction and diffusion, then there will be cause for rejoicing, because a number of analytical tools are available for the qualitative and quantitative treatment of their solutions. Some of them will be taken up in the chapters to follow.

This brings up the subject of qualitative models. Every mathematical model has numerical parameters (rates and probabilities of various occurrences, coefficients in differential equations, the dimensions of spaces, etc.). Even functional, rather than numerical, parameters are common. It is one thing to set up a general model as a framework, and another thing to fill in the parameters. In biological situations, parameters are typically difficult to measure in specific instances, but even without them, some qualitative conclusions can sometimes be deduced. For example, with the bistable nonlinear diffusion equation treated in Chapter 4, very general assumptions on the nature of the function appearing in the equation lead to the existence of stable traveling waves. And for systems of reaction-diffusion equations with an equilibrium point, quite general conditions lead to the bifurcation of small amplitude wave trains.

Our approach in this chapter will be, first, to describe hypotheses under which a phenomenon can be modeled by a system of reaction-diffusion equations. This leaves to be answered the question of which coefficients and reaction functions are the appropriate ones for a given situation. Some discussion of this aspect of modeling will be given in the later sections of the chapter. In the following chapters, the emphasis will be on methods of qualitative analysis of the equations; that is, on discovering conditions under which solutions with certain given properties exist.

In mathematical population biology, qualitative results for models are the most important kind, for accurate quantitative results can only occasionally even be expected. The main reason is that models here are much more idealized than in physics and chemistry. Populations are never homogeneous, the environment is never uniform in time or space, and the population is never isolated from other influences. What one can often hope for, however, is some indication as to the effect, within the total picture, of the few factors and influences specifically being accounted for in the model. Other factors will also leave their marks on the behavior of actual populations. The idealization that these other influences do not exist is expressed by one of our two most important assumptions: that of determinism (A2). (But A2 implies more than just that.)

1.1 Basic hypotheses

A population is a collection of individuals, each occupying a position in space which is a function of time. Instead of position in space, position on some other continuum (such as a phenotypic axis or age axis) is often envisaged. In many such alternate situations, the arguments below apply as well. But for simplicity we shall always speak of geographical space over which the individuals are distributed, at any instant of time. Also for simplicity, we shall talk here only about one-dimensional space (so the population lies on a line, or at least any spatial variation is in one direction only). Many-dimensional reaction-diffusion equations are important, and can be derived by a generalization of the process we describe below. The population is supposed to be divided into several classes which we call species.

The first approach we take is to assume, ab initio, that the population's distribution can be approximated by a smooth density function. In Section 1.9 a different starting point, in which the population is thought of as grouped in colonies, is explored.

For the continuum approach, our two most crucial assumptions are that the various species' spatial distributions can be approximated by smooth density functions, and that the evolution of these functions proceeds deterministically. After these two are made, it is surprising how relatively minor the remaining assumptions appear, which are needed to lead to the reaction-diffusion model. We spell out the two mentioned as follows.

A1. To each species i, there corresponds a (spatial) density function $\rho_i(x,t)$ so that the number of individuals of that species located in the interval $a < x < b$ at time t is approximately $\int_a^b \rho_i(x,t)dx$. These density functions ρ_i are twice continuously differentiable in x, and once in t.

A2. The time evolution of the ρ_i proceeds by a deterministic process, in the sense that if any "initial distribution" in the form of bounded functions

$\rho_i^o(x) \in C^2$ is given at any time t_o, then there is a uniquely determined set of density functions $\rho_i(x,t)$ defined for $t \geq t_o$, with $\rho_i(x,t_o) = \rho_i^o(x)$.

These assumptions are most reasonable when the population is large, as is usually the case in chemical kinetics, where we are concerned with densities of molecules and ions. Biological populations, of course, do not appear continuously distributed in space unless, perhaps, we use a spatial scale which is large compared to the size and spacing of the individuals. Then we may compare the population with a hypothetical "smeared out" substance.

Again, it is sometimes the case that time is most naturally thought of as being discrete, because significant measurements (of the size of the adult population, the seed production, etc.) can only be made a generation apart. If time, then, in effect proceeds in discrete increments, then for a continuous time model to make good sense, we may want the increment size to be small relative to the time scale we are using in the model. Thus, the time unit would be many generations long. Then we hope that the variation on this larger time scale is gradual enough to be capable of approximation by a continuously differentiable function of time. If generations overlap and breeding, etc., is not seasonal, then this particular difficulty does not arise, because at every instant of time there will be individuals in every stage of maturity. But then it may (or may not) be appropriate to distinguish individuals at these various stages; this can be done by labeling them as different species.

The deterministic assumption A2, of course, also is most reasonable when the population is large. Even if one conceives of the world as deterministic, the behavior of each biological organism is subject to so many influences not to be incorporated in the model, that it should be considered largely a matter of chance. Given that it is a matter of chance and supposing the possible behaviors have a definite probability distribution, the actual frequency distribution in behaviors of a given population will most closely reflect the probability distribution, hence be predictable most accurately, when the population is large. Of course even under A2, the density function for a given species, divided by the total population of the species (assuming it is finite), is the probability distribution for position of a random individual of that species. But we are assuming the distribution evolves deterministically.

Deterministic modeling in continuous time is often done even when the population is not very large or the time unit many generations long. The language in the preceding three paragraphs was not meant to convey that such modeling is incorrect. But in such situations, making assumptions A1 and A2 may require a more experienced intuition or a stronger act of faith that the resulting model will qualitatively (at least) reflect the phenomena at hand.

1.2 Redistribution processes

Proceeding further with our assumptions, we shall postulate a migration rule and a reaction (or interaction among the species) rule. The migration rule is best described at first under the condition that no interactions take place. This means no individuals are lost or gained - they only move around. We do this in the present section; in Section 1.4 we shall attach, in very general terms, a "reaction" rule to the migration rule.

With no creation or annihilation of individuals, the migration process will be called a "redistribution" process. Since reactions are absent, we may as well describe it in the case when only one species is present. Our assumptions about this special process will be labeled "R". We assume now the domain of the population is the whole line, relegating the effect of boundaries to the following section.

R1. There is a nonnegative function $g(x,t,y,s)$ $(t > s)$ such that the density satisfies, for any $t > s$,

$$\rho(x,t) = \int_{-\infty}^{\infty} g(x,t,y,s)\rho(y,s)dy. \tag{1.1}$$

For this to make sense for every smooth bounded function $\rho(y,s)$, g should be integrable in y:

$$\int_{-\infty}^{\infty} g(x,t,y,s)dy < \infty. \tag{1.2}$$

This assumption simply supplies more information about the deterministic process spoken of in A2. The two important points about it are

i) the redistribution process is linear, and

ii) for fixed x, t, and s, $\rho(x,t)$ depends continuously on $\rho(\cdot,s)$ in the C^o sense. This means:

Given any $\varepsilon > 0$, there exists a $\delta(\varepsilon)$ (also depending on x, t, s) such that if ρ^1 and ρ^2 are any two density functions satisfying (1.1) and

$$\sup_y |\rho^1(y,s) - \rho^2(y,s)| < \delta, \quad \text{then} \quad |\rho^1(x,t) - \rho^2(x,t)| < \varepsilon. \tag{1.3}$$

This is immediately clear, using (1.2). Thus, small changes in the value of ρ at any one time s induce small changes in ρ at any fixed later time t. Since the effect of individuals at position y, time s, presumably extends in the main only a finite distance during the time interval $[s,t]$, small uniform changes in $\rho(\cdot,s)$ mean small changes in the number of individuals effectively influencing the change at (x,t). This is reasonable.

We could, in fact, generalize R1 to contain only (i) and (ii):

R1'. For any (x,t) and $s < t$, the number $\rho(x,t)$ is a continuous linear functional of the density $\rho(\cdot,s) \in C^o(-\infty,\infty)$.

It is known (see, for example, Riesz-Nagy (1955)) that such functionals may be represented as Stieltjes integrals:

$$\rho(x,t) = \int_{-\infty}^{\infty} \rho(y,s)\,d\alpha_{x,t,s}(y)$$

for some (in this case, bounded nondecreasing) function α. The conclusions below may be reached under this assumption, which is weaker than R1. We leave the argument as an exercise for the reader.

For finite populations, we wish the population size to remain constant. This is the reason for our next assumption:

R2. $$\int_{-\infty}^{\infty} g(x,t,y,s)dx = 1 \quad \text{for} \quad t > s.$$

This assumption emphasizes that g can be interpreted as the probability distribution for the position x (at time t) of an individual located at y at time s. This can be seen by observing that for any two intervals I and J,

$$\int_I dx \int_J g(x,t,y,s)\rho(y,s)dy / \int_J \rho(y,s)dy$$

represents the fraction of those individuals in J at time s which will be in I at time t. The interpretation mentioned comes about when we let J shrink to a point; but the value of the fraction then approaches (for continuous g) the value $\int_I g(x,t,y,s)dx$.

Rather than proceeding as above with R1 and R2, we could alternatively have introduced g initially as this probability distribution. Then R2 would be immediate, but R1, strictly speaking, would be incorrect; (1.1) would yield only the expected value of $\rho(x,t)$, which in fact clashes with A2. This inconsistency could then be mended by saying, in place of A2, that the process is approximately deterministic, and that (1.1) holds approximately. This is presumably reasonable for large populations. We shall meet the problem of inconsistencies among our assumptions again shortly.

Finally, we assume that individuals cannot migrate finite distances in very short time periods. The specific hypothesis is

R3. For every $a > 0$,

$$\lim_{t \downarrow s} \frac{1}{t - s} \int_{|x-y|>a} g(x,t,y,s)dx = 0.$$

This assumption could be interpreted that the probability of an individual's migrating a distance larger than a in a time interval $h = t - s$ approaches zero as $h \to 0$, at a rate faster than h itself. This hypothesis is strange, because it appears to be too <u>weak</u> for our purposes. If one grants that an individual can migrate with only a finite velocity, then it necessarily follows that the probability of one moving a distance $> a$ is zero for small enough time intervals $(t - s)$. Then the integral in R3 will vanish for small $t - s$, and a fortiori R3 will be satisfied. It turns out, however, that if we replaced R3 by the stronger assumption that the integral vanish for small $(t - s)$, and keep the other assumptions, then the implication would be that every individual at (y,s) must move in a predetermined path, and g is a δ-function. If we wish to retain some elements of randomness, or require g to be a genuine function, then we must forego the strict finite velocity concept. This intuitive paradox can be resolved by observing that we expect the model to produce only approximate results anyway, because ρ is only meant to be an approximate density function. We broke with strict reality the moment we committed ourselves to a smooth density function ρ. It turns out that our final result will, in fact, closely approximate the finite velocity situation, because $\int_{|x-y|>a} g dx$ will be extremely small for small $t - s$; in fact, it approaches zero exponentially fast as $t - s \to 0$. R3 is apparently the weakest assumption which will give us what we need. Assumptions like it are common for Markov processes on the line (Feller 1966).

Assumptions A1, A2 and R1-R3 constitute a reasonable model in many cases, but give us apparently little to work with as far as analyzing the dynamics of ρ. The value of the following theorem is to show that they actually provide us with a great deal of information about the dynamics. They imply that ρ is governed by a parabolic differential equation, and that can be of great help in deducing its properties.

In the theory of stochastic processes, it is a common occurrence that the transition probability density function satisfies a parabolic differential equation (see, e.g., Feller (1966)). This type of result was first obtained by Kolmogorov (1931); see also Feller (1936). Nagylaki (1975) applied the reasoning to population problems. The assumptions we make on the evolution process under consideration are to some extent analogous, but are different.

THEOREM 1.1. *To each single-species evolution process subject to* A1, A2, R1-R3, *there correspond functions* $D(x,t) \geq 0$, $C(x,t)$ *such that every density function* ρ *for the process satisfies*

Forward eqn

$$\frac{\partial\rho}{\partial t} = \frac{\partial}{\partial x}\left(D\frac{\partial\rho}{\partial x} + C\rho\right). \tag{1.4}$$

Proof: Consider a density function ρ corresponding to such a process. Let (a,b) be some interval, and let $N(t) = \int_a^b \rho(x,t)dx$ be the content of that interval, as a function of t. Define

$$k(\xi,y,t,h) \equiv g(y + \xi,t + h,y,t) \quad (h > 0),$$

so that

$$\rho(x,t + h) = \int k(x - y,y,t,h)\rho(y,t)dy$$
$$= \int k(\xi,x - \xi,t,h)\rho(x - \xi,t)d\xi.$$

From R2, we have

$$1 = \int k(\xi,y,t,h)d\xi,$$

so that

$$\frac{1}{h}(N(t + h) - N(t)) = \frac{1}{h}\int_a^b dx \int [k(\xi,x - \xi,t,h)\rho(x - \xi,t) - k(\xi,x,t,h)\rho(x,t)]d\xi$$
$$\equiv D_\varepsilon(t,h) + R_\varepsilon(t,h), \tag{1.5}$$

where for any $\varepsilon > 0$

$$D_\varepsilon(t,h) = \frac{1}{h}\int_a^b dx \int_{|\xi|<\varepsilon} [k(\xi,x - \xi,t,h)\rho(x - \xi,t) - k(\xi,x,t,h)\rho(x,t)]d\xi \tag{1.6}$$

and R_ε is the remainder. It follows from R3 that for each $\varepsilon > 0$,

$$\lim_{h\to 0} R_\varepsilon(t,h) = 0. \tag{1.7}$$

By A1, ρ_t is continuous, so $\lim_{h\to 0}(D_\varepsilon + R_\varepsilon) = \int_a^b \rho_t(x,t)dx$; by this and (1.7), we know that D_ε approaches this same integral as $h \to 0$.

By interchanging the variables of integration, we may write

$$\begin{aligned}
D_\varepsilon(t,h) &= \frac{1}{h}\int_{-\varepsilon}^{\varepsilon} d\xi \int_a^b [k(\xi,x-\xi,t,h)\rho(x-\xi,t) - k(\xi,x,t,h)\rho(x,t)]dx \\
&= \frac{1}{h}\int_{-\varepsilon}^{\varepsilon} d\xi \left[\int_{a-\xi}^{b-\xi} - \int_a^b\right] k(\xi,z,t,h)\rho(z,t)dz \\
&= \frac{1}{h}\int_{-\varepsilon}^{\varepsilon} d\xi \left[\int_{a-\xi}^{a} - \int_{b-\xi}^{b}\right] k(\xi,z,t,h)\rho(z,t)dz \\
&\equiv H(a,t,h) - H(b,t,h),
\end{aligned}$$

where

$$\begin{aligned}
H(x,t,h) &= \frac{1}{h}\int_{-\varepsilon}^{\varepsilon} d\xi \int_{x-\xi}^{x} k(\xi,z,t,h)\rho(z,t)dz \\
&= \frac{1}{h}\int_{-\varepsilon}^{\varepsilon} d\xi \int_0^{\xi} k(\xi,x-s,t,h)\rho(x-s,t)ds \\
&= \frac{1}{h}\int_{-\varepsilon}^{\varepsilon} d\xi \int_0^{\xi} k(\xi,x-s,t,h)[\rho(x,t) - s(\rho_x(x,t) + \delta\rho_x)]ds,
\end{aligned}$$

where $\delta\rho_x = \rho_x(\hat{x},t) - \rho_x(x,t)$, $\hat{x} = \hat{x}(s)$ being a value between x and $x - s$. Corresponding to this Taylor expansion for ρ, we have a three-term expression for H:

$$H(x,t,h) = A(x,t,h)\rho(x,t) + B(x,t,h)\rho_x(x,t) + Q(x,t,h). \tag{1.8}$$

First consider the case when $\rho \equiv 0$ for x in a neighborhood of a at time t and $\rho \equiv \rho_o = \text{const}$ in a neighborhood of $x = b$. A2 allows consideration of this case. Then

$$D_\varepsilon(t,h) = -H(b,t,h) = -A(b,t,h)\rho_o,$$

and since this approaches a limit as $h \to 0$, we have

$$C(x,t) \equiv -\lim_{h\to 0} A(x,t,h)$$

exists for all x and t. Next, suppose $\rho(x,t)$ is linear in an ε-neighborhood of b. Then since $\delta\rho_x = 0$, $D_\varepsilon(t,h) = -A\rho - B(x,t,h)\rho_x$, and we conclude that

$$-D(x,t) \equiv \lim_{h\to 0} B(x,t,h)$$

exists.

Now

$$|Q(x,t,h)| \le \frac{1}{h}\int_{-\varepsilon}^{\varepsilon} d\xi \int_0^{\xi} sk(\xi, x - s, t, t + h)|\delta\rho_x| ds$$

$$\le \sup_{|s|<\varepsilon} |\delta\rho_x| B(x,t,h),$$

so

$$\limsup_{h\to 0} |Q| \le \delta(\varepsilon) D(x,t),$$

where by continuity of ρ_x, $\delta(\varepsilon) \to 0$ as $\varepsilon \to 0$.

But from the above, we know that $\lim_{h\to 0} Q(x,t,h)$ exists and is independent of ε, so this limit must be zero. Putting all this together, we have

$$\int_a^b \rho_t dx = \lim_{h\to 0} \frac{N(t + h) - N(t)}{h} = H(a,t) - H(b,t), \tag{1.9}$$

where

$$H(x,t) \equiv \lim_{h\to 0} H(x,t,h) = -C(x,t)\rho(x,t) - D(x,t)\rho_x(x,t).$$

The left side of (1.9) may be differentiated with respect to the upper limit b. This shows H also to be differentiable with respect to x, and

$$\rho_t = (D\rho_x + C\rho)_x, \tag{1.10}$$

which is what we wished to prove.

It follows from (1.9) that $H(x,t)$ represents the flux, i.e., the number of individuals per unit time crossing the position x from left to right. The fact that the flux is a linear function of ρ_x and ρ is sometimes called Fick's law. It also follows from (1.10) that stationary density configurations are characterized by the condition $H(x,t) = \text{const}$.

From its definition, clearly $D \geq 0$. If it should happen that $D \equiv 0$ is some open set in the (x,t) plane, then (1.10) is a first order equation with well-defined characteristics in that region. These lines are the required paths of the individuals of the population. If the portion of the characteristic through (x,t) extending down to the line "time = s" lies entirely within the open set, then the function $\alpha_{x,t,h}$ in R1' is constant except for a jump discontinuity. To admit this case, R1' would have to be used in place of R1 (unless one accepts δ-functions as functions in the statement of R1).

Age structured models Weiss, Gopalsamy

For $D > 0$, solutions of (1.10) may be represented in the form (1.1) with g the "fundamental solution" for the equation (see, for example, Friedman (1964)). Much is known about such fundamental solutions, and especially about the singularity as $(x,t) \to (y,s)$. In the case of a homogeneous time-invariant environment (D and C constant), we have the explicit formula

$$k(\xi,y,t,h) = \frac{1}{\sqrt{4\pi Dh}}\, e^{-\frac{(\xi+Ch)^2}{4Dh}} .$$

An example of an easy consequence of this is the following condition, much stronger than R3:

$$\lim_{t\downarrow s} \frac{1}{(t-s)^N} \int_{|x-y|>a} g(x,t,y,s)\,dx = 0 \quad \text{for all } N \text{ and } a > 0.$$

The few seemingly minor assumptions about the evolution process being considered are thus seen to circumscribe it to a surprising degree.

The coefficient D can be related to the concept of "infinitesimal variance" in migration rate as follows. This latter is defined to be

$$V(s,t) = \lim_{h\to 0} \frac{1}{h} \int_{-\infty}^{\infty} \xi^2 k(\xi,x,t,h)d\xi. \tag{1.11}$$

Similarly, the "infinitesimal drift" is defined to be

$$M(x,t) = \lim_{h\to 0} \frac{1}{h} \int_{-\infty}^{\infty} \xi k(\xi,x,t,h)d\xi. \tag{1.12}$$

Recall that

$$-B(x,t,h) = \frac{1}{h} \int_{-\varepsilon}^{\varepsilon} d\xi \int_{0}^{\xi} sk(\xi,x - s,t,h)ds.$$

Let us assume that k is continuous in its second argument with modulus of continuity bounded by a multiple of k itself, in the sense that

$$k(\xi,x - s,t,h) = k(\xi,x,t,h)(1 + q),$$

where $q \to 0$ as $s \to 0$. Then

$$-B(x,t,h) = \frac{1}{2h} \int_{-\varepsilon}^{\varepsilon} \xi^2 k(\xi,x,t)d\xi(1 + \bar{q}),$$

where $\bar{q} \to 0$ as $\varepsilon \to 0$. From R3 and (1.11), we see that

$$D(x,t) = -\lim_{h\to 0} B(x,t,h) = \frac{1}{2} V(x,t).$$

In a similar manner, if we assume that k_x exists and has modulus of continuity bounded by a multiple of k, and also assume

$$-\lim_{h\to 0} B_x(x,t,h) = D_x(x,t) = \frac{1}{2} V_x(x,t),$$

we obtain

$$C(x,t) = -M(x,t) + \frac{1}{2} V_x(x,t).$$

Then (1.10) becomes

$$\rho_t = \left(\frac{1}{2} (V\rho)_x - M\rho \right)_x . \tag{1.13}$$

This is called Kolmogoroff's forward equation. It was first derived by him (Kolmogoroff 1930) under assumptions different from those listed above, for the transition probability function for a time-continuous stochastic process on the real line.

In the following, we shall refer to the function $g(x,t,y,s)$ appearing in (1.1) for a redistribution process satisfying A1, A2, R1-R3 as a "redistribution kernel." As mentioned above, it is simply a fundamental solution for the diffusion equation (1.4).

1.3 Boundaries and interfaces

Although our model began with a global description (1.1) of the dynamics of ρ, the assumptions led to a differential equation (1.4), which implies the dynamics are governed by local events. Therefore when we examine the effect of boundaries, the first thing to realize is that (1.4) continues to hold in the interior of the domain, where local events do not include boundary events.

At the boundary of the domain, however, the migration process described before is disrupted in some manner. A complete description of the dynamics of a population must include a specification of the nature of this disruption. Various possibilities suggest themselves:

(1) Individuals are prevented from exiting or entering through the boundary. This is the "no flux" condition

$$H(x_o,t) = 0,$$

where x_o is a boundary point (in the following, we take it to be the right-hand boundary point, to be precise.)

(2) Individuals are imported or exported at a given rate:

$$H(x_o,t) = a(t).$$

Then, of course, it will no longer be true that the total population size remains constant, and so (1.1) will not hold (even if the integration domain is restricted) with a kernel satisfying R2. Nevertheless, as mentioned above, the same local process (1.4) proceeds as before, except on the boundary.

(3) The efflux through the boundary is a linear function of ρ at that point:

$$H(x_o,t) = b(t)\rho(x_o,t) - a(t).$$

This is sometimes referred to as the "radiation," "Robin," or "third" boundary condition. An example of its use is when the boundary is a permeable membrane, with flux across it proportional to the difference in concentrations (densities) between the two sides.

(4) The "Dirichlet" condition, in which ρ is prescribed:

$$\rho(x_o,t) = c(t).$$

This is the limiting case of (3) when a and $b \to \infty$ with $c = a/b$ remaining finite. The membrane then would be highly permeable.

Not only does it make sense physically to specify the boundary conditions in addition to the migration rule, but we must do so to obtain well-posed mathematical problems. More about this in Chapter 3.

An interface is a location within the domain of ρ where the migration process is disrupted. Typical causes for disruption may be a membrane partitioning a chemical mixture, or a geographical barrier (river, etc.) to migration of a biological population. The type of disruption, again, has to be specified by an extra side condition at the interface. One typical side condition (Slatkin 1973; Nagylaki 1976) is that the flux through the barrier is proportional to the discontinuity in ρ:

$$-H(x_o,t) = b(t)[\rho(x_o + 0,t) - \rho(x_o - 0,t)],$$

the flux itself remaining continuous.

An abrupt change in the properties of the habitat at one point could also be considered an interface effect. In this case, the domain of ρ would be separated into the section $x < x_o$, say, and $x > x_o$, with the quantities D and C from (1.10) discontinuous at $x = x_o$ (Nagylaki 1976). Then since no barrier is involved, ρ would be continuous at x_o, and the obvious balance law would require the flux also to be continuous:

$$H(x_o + 0,t) = H(x_o - 0,t),$$

where $H = C\rho + D\rho_x$ as usual. Since C and D are discontinuous, this will result in ρ_x being discontinuous at x_o.

1.4 Reactions with migration

The foregoing sections were about continuum models of a population whose individuals can move around, but cannot die or beget new ones. The following is a way to construct a general continuum model which includes the creation and annihilation of individuals. In chemically reacting mixtures, new molecules appear at the expense of others (which may have combined or decomposed to produce the new ones). And in biological populations, reaction effects include predation, reproduction, and natural death. In either case, the population should be differentiated into various species (chemical or other), age groups, genotypes, etc. (all of which we shall also call species). So in the following, we speak of a population of n species or classes, and denote the vector of densities of these species by $\underline{\rho} = (\rho_1, \ldots, \rho_n)$.

I shall first attempt to motivate the assumptions to be made, and afterwards list them formally.

Use will be made of redistribution kernels, which were formally defined on page 17. But, in place of a single such kernel $k(\xi,y,t,h) = g(y + \xi,t + h,y,t)$ we now have one for each species: $\underline{k} = (k_1, \ldots, k_n)$. If no deaths or births were allowed, the dynamics would be governed by

$$\rho_i(x,t+h) = \int k_i(x-y,y,t,h)\rho_i(y,t)dy,$$

or, denoting the integral operators by $K_i(t,h)$ for short,

$$\rho_i(\cdot,t+h) = K_i(t,h)\rho_i(\cdot,t).$$

In a similar way, we now let $\underline{v}(x,t,h)$ denote the vector densities, at time $t+h$, of new individuals which have appeared in the population during the time interval t to $t+h$. We don't as yet spell out how they make their appearance. We wish to have the function $\underline{v}$ account also for the disappearance of individuals; such occurrences will give a negative contribution to $\underline{v}$, and in fact the v_i themselves may be negative or positive. A negative density $v_i(x,t,h)$ of (vanished) individuals would simply be equal in magnitude to the contribution to $\rho_i(x,t+h)$, at the same location and time, that these individuals would have made had they not vanished. So in a sense, dead individuals are thought of as migrating according to the same law as live ones, but produce a negative density.

According to the above, our model is now

$$\rho_i(\cdot,t+h) = K_i(t,h)\rho_i(\cdot,t) + v_i(\cdot,t,h), \quad h > 0. \tag{1.14}$$

The first basic assumption on $\underline{v}(x,t,h)$ is that it is determined, once the vector function $\underline{\rho}(\cdot,t)$ is known. We symbolize this by the relation

$$\underline{v}(x,t,h) = h\underline{F}(x,t,h,[\underline{\rho}(\cdot,t)]), \tag{1.15}$$

where $[\underline{\rho}]$ means that $\underline{F}$ depends on the entire function $\rho(\cdot,t)$, not on its values at particular points. The functional $\underline{F}$ will, of course, generally be nonlinear. It will turn out that $\underline{F}$ is bounded in h, so that (1.15) implies $\underline{v}(x,t,0) = 0$; intuitively this should be the case.

Our second basic assumption is a rather minimal one about the dependence of $\underline{F}$ on $\underline{\rho}$. We want to say that the vector $\underline{v}$ at a point x will not depend very much on the value of $\underline{\rho}$ at distant points, because the individuals at the latter points, and their reaction products, will not have had sufficient time to migrate to x. In accounting for this dependence, we therefore propose a "weight" function $\bar{k}(\xi,h)$ which will be a crude measure for the relative influence exerted by the value of $\underline{\rho}(x+\xi,t)$ on the value of $\underline{v}(x,t,h)$. It ought to have the main properties of the migration kernels k_i, but to be conservative we only require that

$$\bar{k} \geq 0, \quad \int_{\infty}^{\infty} \bar{k}(\xi,h)d\xi < C, \quad \text{(independently of } h) \tag{1.16}$$

$$\lim_{h\to 0} \int_{-\infty}^{\infty} \xi^2 \bar{k}(\xi,h)d\xi = 0. \tag{1.17}$$

Our assumption, now, is that

> for each fixed (x,t,h), $\underline{F}$ is a functional on $L_1^{(\bar{k})}$ (L_1 with weight function $\bar{k}$) which is continuous, uniformly in h. (1.18)

In other words, for every (x,t) and $\varepsilon > 0$, there is a $\delta(x,t,\varepsilon)$ such that if $\underline{\rho}^{(1)}$ and $\underline{\rho}^{(2)}$ are two functions of x with

$$\int_{-\infty}^{\infty} \bar{k}(x-y,h)|\underline{\rho}^1(y) - \underline{\rho}^2(y)|dy < \delta,$$

then

$$|\underline{F}(x,t,h,[\underline{\rho}^{(1)}]) - \underline{F}(x,t,h,[\underline{\rho}^{(2)}])| < \varepsilon$$

for all $h > 0$.

LEMMA 1.2. Assume (1.18) holds, with (1.16) and (1.17). Let $\rho^{(1)}$ and $\rho^{(2)}$ be two functions in $C_2(-\infty,\infty)$ with $\rho^{(1)}(x_o) = \rho^{(2)}(x_o)$ for some x_o. Then

$$\lim_{h\to 0} |\underline{F}(x_o,t,h,[\rho^{(1)}]) - \underline{F}(x_o,t,h,[\rho^{(2)}])| = 0.$$

<u>Proof</u>: By Taylor expansion,

$$\rho_i^{(1)}(y) - \rho_i^{(2)}(y) = (\rho_{ix}^{(1)}(x_o) - \rho_{ix}^{(2)}(x_o))(y - x_o)$$

$$+ \frac{1}{2}(\rho_{ixx}^{(1)}(\hat{x}) - \rho_{ixx}^{(2)}(\hat{x}))(y - x_o)^2,$$

where $\hat{x}$ is between x_o and y. Thus

$$\left| \int_{-\infty}^{\infty} \bar{k}(x - y,h)[\rho_i^{(1)}(y,t) - \rho_i^{(2)}(y,t)]dy \right| \leq$$

$$\leq C\left(\int_{-\infty}^{\infty} |x - y|\bar{k}(x - y,h)dy + \int_{-\infty}^{\infty} (x - y)^2\bar{k}(x - y,h)dy\right). \quad (1.19)$$

But by the Schwarz inequality,

$$\int |y - x|\bar{k}(x - y,h)dy = \int |\xi|\bar{k}(\xi,h)d\xi$$

$$\leq \left(\int \bar{k}(\xi,h)d\xi\right)^{1/2}\left(\int \xi^2\bar{k}(\xi,h)d\xi\right)^{1/2} \to 0$$

as $h \to 0$.

Similarly, the last integral in (1.19) vanishes as $h \to 0$. This, together with the continuity property (1.18), proves the lemma.

We are now ready to state the formal assumption.

HYPOTHESIS. There are redistribution kernels $k_i(x,t,y,s)$ $(t > s)$ and a functional $\underline{F}(x,t,h,[\underline{\rho}])$ defined for all x, t, $h > 0$ and $\underline{\rho} \in C_o(-\infty,\infty)$, and satisfying (1.18), (1.16), (1.17), such that for every x, t, and $h > 0$,

$$\rho_i(x,t + h) = \int_{-\infty}^{\infty} k_i(\xi,x - \xi,t,h)\rho_i(x - \xi,t)d\xi + hF_i(x,t,h,[\rho(\cdot,t)]).$$

The object now will be to show that this hypothesis implies that $\underline{\rho}$ is governed by a RD system.

THEOREM 1.3. To each evolution process satisfying the above hypothesis, there correspond functions $D_i(x,t) \geq 0$, $C_i(x,t)$, $F_i(x,t,\underline{u},)$ such that every density function $\underline{\rho}(x,t)$ for the process which is continuously differentiable in t and twice continuously differentiable in x satisfies[†]

$$\frac{\partial \rho_i}{\partial t} = \frac{\partial}{\partial x}(D_i \frac{\partial \rho_i}{\partial x} + C_i\rho_i) + F_i(x,t,\underline{\rho}(x,t)). \tag{1.20}$$

<u>Proof</u>: On the basis of the proof of Theorem 1.1, we know that for each (x_o, t_o),

$$\lim_{h\to 0} \frac{1}{h}\left((K_i(t_o,h)\rho_i\Big|_{\substack{x=x_o\\t=t_o}} - \rho_i(x_o,t_o)\right) = \frac{\partial}{\partial x}(D_i(x,t)\frac{\partial\rho_i}{\partial x} + C_i(x,t)\rho_i)_{x=x_o,t=t_o}$$

$$\equiv L_i\rho_i(x_o,t_o),$$

for some coefficients C_i and D_i. This, together with our hypothesis, tells us that the limit $\lim_{h\to 0} \mathcal{F}_i(x_o,t_o,h,[\underline{\rho}(\cdot,t_o)])$ exists, and

$$\frac{\partial}{\partial t}\rho_i(x_o,t_o) = L_i\rho_i(x_o,t_o) + \mathcal{F}_i(x_o,t_o,0,[\underline{\rho}(\cdot,t_o)]),$$

where the last term denotes this limit. Lemma 1.2 implies that the limit depends only on the value $\underline{\rho}(x_o,t_o)$ rather than on the entire function $\underline{\rho}(\cdot,t_o)$. Thus

$$\mathcal{F}_i(x,t,0,[\underline{\rho}(\cdot,t)]) \equiv F_i(x,t,\underline{\rho}(x,t)).$$

This completes the proof.

Note that if the environment is homogeneous in space and time, the D_i, C_i, F_i will not depend on x and t. In that case, we have

$$\underline{\rho}_t = D\underline{\rho}_{xx} + C\underline{\rho}_x + \underline{F}(\underline{\rho}),$$

for a vector function $\underline{F}$, where D and C are diagonal matrices, D with non-negative elements.

[†]A criticism made by Mollison (1976), Section 0.1, is relevant here.

1.5 The reaction mechanism

Since $\underline{v}(x,t,h)$ represents the amount (density) of new individuals produced in the time interval from t to $t + h$, and the function $\underline{F}$ in (1.20) is defined by

$$\underline{F}(x,t,\underline{\rho}(x,t)) \equiv \lim_{h\to 0} \frac{1}{h}\,\underline{v}(x,t,h),$$

this function may be called the rate of production due to reaction, or source function. The usual approach in determining the appropriate source function is to envisage a reaction mechanism. It is appropriate to emphasize at this point, however, that the diffusing and/or reacting quantity being modeled is often not a population at all, but a physical or thermodynamic property of a medium, the medium itself sometimes being a population.

A prime example is heat in a reactor. Since chemical reactions generally produce or absorb it, the time rate of change in temperature depends on the rates of the various chemical reactions. These rates, moreover, depend on the temperature as well as on the concentrations of the reacting species. This situation may be described within the framework of equation (1.20) by letting one of the components of $\underline{\rho}$ denote temperature. One can imagine other states of a population that one might wish to model this way, for example, cultural states or learning levels in a human population, or disease states in a biological population.

The following discussion of reaction mechanisms and related concepts is confined principally to reactions which occur among individuals of a population. It is based mainly on the very readable account given by Feinberg and Horn (1974) and Feinberg (1978). Much of the theory we shall allude to was developed by these authors. The reader is also referred to Vol'pert and Hudyaev (1975), Aris (1965), Gavalas (1968), Feinberg (1972), Feinberg and Horn (1977), and Horn (1972) for development of the theory.

In the case of chemical reactions and usually in other cases as well, this reaction mechanism can be symbolized in the following manner. First, one decides on symbols for the various species. We use the symbols $A_1, \ldots, A_n$, assuming there are n species in all. Then the reactions are symbolized as in the following examples:

$$A_1 \to A_2$$

$$A_1 + A_2 \rightleftarrows A_3 + A_4 \to A_5$$

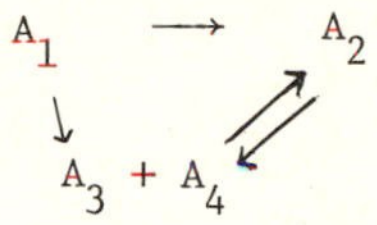

Reactions involving two arrows are called reversible. The arrows represent events which led to the creation or disappearance of individuals. For example, $A_1 + A_2 \to A_3$ may mean that a molecule of A_1 combines with one of A_2 to produce a compound A_3; and $A_3 \to A_1 + A_2$ means that a molecule of A_3 dissociates into the other two species. The event $A_1 \to A_2$ may imply the radioactive or other transformation of a molecule of A_1 into A_2. The birth of a new individual of species A_1 might be denoted by $A_1 \to 2A_1$. Other units besides individuals in a population may be appropriate in visualizing reaction mechanisms. For example, if A_1 is a prey species and A_2 is a predator, the event $A_1 + A_2 \to 2A_2$ signifies that one unit of predator consumes one unit of prey, and the effect is to produce (through a time delay, of course, which is neglected) another unit of predator. If the predator is an anteater, the appropriate question is, how many ants will support the production of one new anteater? Whatever the number is, that would be the appropriate unit of species A_1, if individuals are used as the unit of A_2. An alternative procedure would be to choose, from some other consideration, a convenient unit for A_1, and replace the above reaction by $A_1 + A_2 \to (1 + \alpha)A_2$, where $\alpha < 1$ is the number of new units of A_2 produced by consuming one unit of A_1. However, for some purposes it will be convenient to have the coefficients of the species symbols always a nonnegative integer.

If the possible interactions in the population are exclusively of the type described above, the set of reactions is called a <u>closed system</u>. This means that recombinations and transformations occur only among the species within the system.

Open systems are otherwise, and at least three typical mechanisms may be considered, which do not occur for closed systems:

(i) New individuals of some species are added, at a given rate, to the population. In the case of a chemical reactor, this is often done: new material is simply fed in from the outside. If A_1 is the species added, the event may be symbolized in the fashion

$$0 \to A_1.$$

(ii) Some individuals are taken away from the system at a given rate (which may, and usually does, depend on the concentration of that species). This is symbolized by

$$A_1 \to 0.$$

(iii) An amount of some species is added, at a rate which will result in the concentration of the species being constant. Although a mechanism for doing this is hard to visualize, it is often the case that one species is present in such abundance that, though it takes part in reactions, the relative variation in its numbers resulting therefrom is so small that the concentration may be considered effectively constant. For purposes of deriving the kinetics of a reactor, it is convenient simply to omit all mention of such constant-density species in writing down the various reactions. For example, if species A_2 is constant, the reaction $A_1 + A_2 \to A_3$ can simply be written $A_1 \to A_3$.

The collection of all reactions occurring in a given system (open or closed), symbolized as indicated above, will be called the "reaction mechanism" for the system. It should be noted that many chemical reaction processes observed in the laboratory or in nature have exceedingly elaborate mechanisms. In fact their determination, or even the determination of an approximation to the exact mechanism, is a major problem unsolved in many cases. Here we assume the mechanism is known, or else assume the actual mechanism can be modeled by an explicit known mechanism.

Besides species, it will be important to consider "complexes," which are defined as the expressions occurring at either end of the arrows in the reaction mechanism. Thus if the mechanism is

$$A_1 + A_2 \to 2A_3 \rightleftarrows A_3 + A_4, \quad A_3 \to 0, \tag{1.21}$$

then there will be five complexes: $A_1 + A_2$, $2A_3$, $A_3 + A_4$, A_3, and 0. Let m be the total number of complexes. If a complex occurs more than once in a mechanism, it is counted only once. Complexes may be denoted by vectors in "species space," by which we simply mean $\mathbb{R}^n$ (n-dimensional Euclidean space). For example, in the above mechanism, there are four species $A_1, \ldots, A_4$, so $n = 4$, and each complex is a 4-vector, namely (1,1,0,0), (0,0,2,0), (0,0,1,1), (0,0,1,0), and (0,0,0,0).

We now introduce the reaction vectors for a mechanism. There will be one for each reaction, hence one for each arrow. They are defined in each case as the complex to which the arrow points, minus the complex away from which it points. In the case above, the reaction vectors are (-1,-1,2,0), (0,0,-1,1), (0,0,1,-1), and (0,0,-1,0).

The stoichiometric subspace for a given mechanism is the subspace of species space spanned by the reaction vectors. It will be denoted by S, and its dimension by s. Of course $s \leq m$. In the above example, $s = 3$, because three of the reaction vectors are linearly independent, but two are negatives of each other.

For future reference, we introduce two additional indices for a mechanism. A linkage class is a collection of complexes which can be linked to one another by means of arrows or chains of arrows. Moreover, it is a maximal such collection. In the example (1.21), there are two linkage classes; in general we denote the number of them by ℓ. Finally, we define the "deficiency" of a mechanism by

$$\delta = m - s - \ell. \tag{1.22}$$

It turns out (Feinberg 1972) that $\delta \geq 0$. Furthermore, it so happens that most chemical reaction mechanisms which one encounters have $\delta = 0$. Mechanisms with deficiency 0 often give rise to dynamical behavior of a particularly simple type.

Along with an assumed reaction mechanism, as described above, we also assume given a set of nonnegative functions of x, t, and $\underline{\rho}$ called rate functions. There will be one such function for each reaction. For convenience we shall suppress the dependence on x and t. Let us denote the m complexes by the symbols y_i, $i = 1, \ldots, m$, and by $f_{ij}(\underline{\rho})$ the rate function (see below) for the reaction, if it exists, going from complex j to complex i. If there is no such reaction, then we set $f_{ij} = 0$. Along with these individual rate functions, we define an overall rate of production

$$\underline{F}(\underline{\rho}) = \sum_{i,j}^{n} f_{ij}(\underline{\rho})(y_i - y_j). \tag{1.23}$$

Note that the overall rate function is vector-valued.

The individual function f_{ij} is supposed to give the rate by which the amount of complex i in the system increases, and the amount of complex j decreases, due to the reaction $j \to i$. In making this interpretation, a host of implicit assumptions are also made, some of them mutually contradictory. For example, the transformation of complex j to complex i represented by the reaction $j \to i$, is assumed to be instantaneous, so there are no "intermediate" complexes. And yet these instants are somehow spread out in time, so that it is a process which occurs continuously. But we leave aside these questions, and assume the rate functions are given.

With due regard to the fact that a known rate of change in the concentration of any one complex due to any one reaction induces a corresponding change in the concentration of each species entering into the complex, weighted by its stoichiometric coefficient, it is seen quite easily that the i-th component of $\underline{F}$ represents the total rate of change of ρ_i due to all the reactions combined. It is therefore identical with the source function in (1.20).

The collection of all rate functions is called the "kinetics" associated with the given reaction mechanism. It is clear that $\underline{F}$, being a combination of the reaction vectors, always lies in the stoichiometric subspace S.

1.6 Positivity of the density

Since the function $\underline{\rho}$ is a density vector, its components must be nonnegative. Not all solutions of (1.20) are nonnegative, however, and so this extra restriction must (at least implicitly) be imposed when dealing with that equation. Still, it may happen that nonnegative initial data evolve into solutions of (1.20) which become negative at some later time. If such were the case, the model would lead to nonsense and would have to be revised. This pitfall, it turns out, may be avoided by a very reasonable assumption about the reaction rates:

> The functions f_{ij} are differentiable; and
> $f_{ij}(\underline{\rho}) = 0$ if $\rho_k = 0$ for some k such that
> $y_{j,k} \neq 0.$ (1.24)

Here $y_{j,k}$ denotes the k-th component of the vector y_j, and the condition $y_{j,k} \neq 0$ means that species k enters into complex j. The assumption simply says that the reaction $j \to i$ can proceed only if there is present some of each species composing complex j.

The following positivity result follows. (For more general results, see Auchmuty (1976, **1978**).) Here we denote by V^+ the set of vectors in $\mathbb{R}^n$, each of whose components is positive. Its closure (the set of vectors with nonnegative components) is denoted by $\overline{V}^+$.

THEOREM 1.4. Assume (1.24). Let $\underline{\rho}(x,t)$ be a solution of (1.20) such that $\underline{\rho}(x,0) \in \overline{V}^+$. Then $\underline{\rho}(x,t) \in \overline{V}^+$ for all (x,t) with $t \geq 0$.

Proof: For fixed k, we may write, from (1.24),

$$F_k(\underline{\rho}) = \Sigma_1 f_{ij}(\underline{\rho})(y_{i,k} - y_{j,k}) + \Sigma_2 f_{ij} y_{i,k},$$

where Σ_1 is the summation over all reactions $j \to i$ such that $y_{j,k} \neq 0$, and Σ_2 is over those for which $y_{j,k} = 0$. By (1.24), all the f_{ij} in the first summation, hence the entire first summation, must vanish when $\rho_k = 0$, so it can be represented as $\rho_k a(x,t)$ for some bounded function a. And since $f_{ij} \geq 0$, the second summation is nonnegative. Therefore

$$\frac{\partial \rho_k}{\partial t} - L_k \rho_k = F_k(\underline{\rho}) = \rho_k a(x,t) + b(x,t)$$

with $b \geq 0$ and a bounded. The strong maximum principle for parabolic equations (see, e.g., Friedman (1964), Protter and Weinberger (1967)), now implies that that $\rho_k > 0$ for $t \geq 0$, as required.

If we only assume that $\underline{\rho}(0) \in \overline{V}^+$, then the weak maximum principle implies that $\underline{\rho}(t) \in \overline{V}^+$ for all t, and in fact the only components which could possibly vanish for finite t are those which were zero initially.

1.7 Homogeneous systems

In this and the next section, we speak of a population of more than one species, whose densities do not vary in space. This is the case, for example, with a "continuous stirred tank reactor," studied extensively in chemical engineering. Such a reactor is simply a vessel containing a homogeneous mixture of chemical substances. Chemicals may be added to the mixture, and material withdrawn, but at all times it is kept well mixed, so that spatial homogeneity prevails. The only variation in chemical composition, then, will be a temporal one.

This is a case of induced mobility; other situations which can be modeled this way are those in which the population is intrinsically very mobile. By this we mean that the average time it takes an individual to diffuse or otherwise travel from one end of the habitat to the other will be small compared to the time scale in which "reaction" effects (chemical reactions, deaths, births, disease infections, etc.) serve to change the composition of the population appreciably. So even if the migration rate is not large, it may happen that other changes are even slower, and for this reason that the "stirred tank" approximation is valid.

Finally, if in the system (1.20) none of the functions appearing in the equation depend on x, and initially $\underline{\rho}(x,0) = \text{const}$, then $\underline{\rho}$ will remain x-independent for later times as well.

At any rate, whether the spatial uniformity results from natural or artificial means, or because of uniform initial data, the foregoing analysis simplifies and leads to (1.20) with the x-derivative terms absent:

$$\frac{d\underline{\rho}}{dt} = \underline{F}(t,\underline{\rho}). \tag{1.25}$$

This, the system of kinetic equations, has of course been studied far more extensively than (1.20).

It is a curious fact that the condition opposite from that of extreme mobility can also be modeled by the same dynamical equations as those which serve for the stirred tank reactor. For spatially homogeneous populations, there is no diffusion or drift terms in the dynamical equations, because spatial derivatives vanish. But consider the situation when the time required for an individual to migrate a significant distance is large compared to the time for reaction effects to produce meaningful changes. Then within the reaction time scale, one may neglect spatial movement. This results in the elimination of the diffusion and drift terms from the dynamical equations. In such a situation there may very well be spatial dependence of the concentrations, but this is not seen in the equations, except possibly in the sense that the function F may involve x as a parameter, to reflect an inhomogeneous environment.

This argument for eliminating spatial derivative terms from (1.20) was based on a comparison of typical reaction and migration scales. In such situations there may still be a need to account for the effects of diffusion. Specifically, (1) we may wish to examine the long-term behavior of the dynamics, and so employ a slow time scale, or (2) boundary or internal layers (steep concentration gradients) may occur. These are cases when the characteristic length and time scales are brought into balance again by shortening one or extending the other. Then the diffusion terms in (1.20) will again be important. We shall deal with these effects later.

We assume the function $\underline{F}$ is time-independent, and is built from some reaction mechanism, as described in Section 1.4. It is clear, then, that each trajectory of (1.25) must lie in some parallel of the stoichiometric subspace S.

This is true because $\underline{F} \in S$, so (integrating (1.25))

$$\underline{\rho}(t) - \underline{\rho}(0) = \int_0^t \underline{F}(\underline{\rho}(\tau))d\tau \in S',$$

and $\underline{\rho}(t) \in S + \underline{\rho}(0)$.

In the last section, it was shown that the densities remain positive if they start out that way.

Another property which in a great many cases is obvious for the phenomenon, and which we would want to be reflected in the model, is the property of boundedness. For example, a closed isothermal chemical system is such that all concentrations must be bounded functions of time. This is clear because no new molecules are added to the system. Of course, each individual species is not conserved, because chemical reactions take place. But the number of atoms, for example, would be conserved, and this leads to an upper bound for the concentration of any one species.

Boundedness is an example of containment properties, to which we shall return later in Chapter 5. But now we show that the stirred tank reactor model does exhibit this boundedness.

Let $p_j > 0$ be the number of atoms in species A_j. Then setting $p = (p_1, \ldots, p_m)$, and y_k the vector representing the k-th complex, we see that the scalar product (p,y_k) is the number of atoms in complex k. Conservation of atoms simply means that $(p,y_k) = (p,y_j)$ for any pair of complexes joined by a reaction. Hence p is orthogonal to every vector in S, and therefore to every stoichiometric parallel. Since $p \in V^+$, it easily follows that each such parallel has a bounded intersection with $\bar{V}^+$. But each orbit must lie in such an intersection, and so must be bounded.

A similar argument holds, of course, if there is dissipation of atoms (such as by leaking through a membrane), and in fact we have the following general boundedness criterion:

THEOREM 1.5. Suppose there exists a vector $p \in V^+$ such that $(p,r) \leq 0$ for every reaction vector r. Then each orbit is bounded.

In case we know a priori (as in this theorem) that each potential orbit is bounded, we also have a global existence theorem for solutions of (1.25). A similar result holds for (1.20).

1.8 Modeling the rate functions

The kinetics of chemical reactors is often taken to be of mass action type. This means that the rate of each reaction is proportional to the concentration of each species in the complex entering the reaction, raised to the power equal to its stoichiometric coefficient. One rationale behind this type of kinetics is that the event symbolized by the given reaction occurs only when individuals in the entering complex encounter one another, and the rate of these encounters is proportional to the concentration of each species in the encounter. Again, this reasoning can sometimes be justified on the basis of kinetic theory, and for low density populations is intuitively reasonable.

The law of mass action, then, takes the form

$$f_{ij}(\underline{\rho}) = k_{ij}\underline{\rho}^{y_i}, \tag{1.26}$$

where k_{ij} is a constant, the "rate of constant" for the reaction $j \to i$, and

$$\underline{\rho}^{y_j} \equiv \prod_{\ell=1}^{m} \rho_\ell^{y_{j,\ell}} . \tag{1.27}$$

In the case of the reaction

$$0 \to A_1,$$

(1.26) reduces to $f_{ij} = k_{ij}$, representing an influx rate independent of any concentration.

Mass action kinetics, of course, satisfies (1.24). One obvious modification would be when ρ_m represents temperature; then the product in (1.27) should be replaced by one in which the upper limit is $m - 1$, and in (1.26) one should allow

k_{ij} to depend on temperature.

Example 1:

$$2A_1 \underset{k_{12}}{\overset{k_{21}}{\rightleftarrows}} A_2.$$

Here mass action kinetics with rate constants k_{12}, k_{21}, applied to (1.25), gives

$$\dot{\rho}_1 = -2k_{21}\rho_1^2 + 2k_{12}\rho_2$$

$$\dot{\rho}_2 = k_{21}\rho_1^2 - k_{12}\rho_2.$$

The stoichiometric subspace is spanned by the single vector $(-2,1)$, so $s = 1$. Furthermore, it is perpendicular to $p = (1,2) \in V^+$, so by Theorem 1.5, all orbits are bounded and lie on the parallels indicated below.

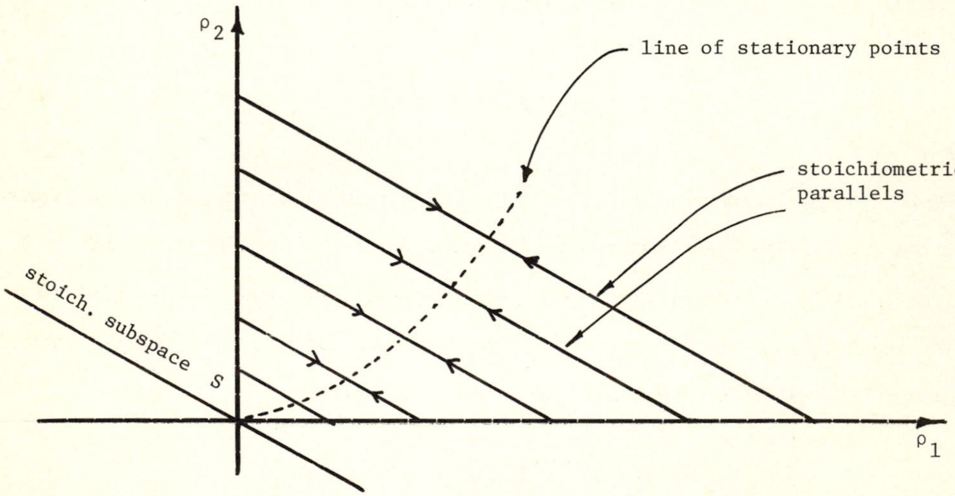

Stationary points are values (ρ_1,ρ_2) satisfying $h(\rho_1,\rho_2) \equiv k_{12}\rho_2 - k_{21}\rho_1^2 = 0$. There is exactly one such point on each parallel (restricted to V^+ of course). They are on the dotted parabola indicated. Inside the parabola, $h(\rho_1,\rho_2) > 0$, so we have $\dot{\rho}_1 > 0$ and $\dot{\rho}_2 < 0$, whereas the opposite inequalities hold outside. The orbits are therefore directed as indicated by the arrows. It follows that each such stationary point is globally stable, in the sense that an orbit starting anywhere on that parallel in V^+ will approach it as $t \to \infty$.

We compute the various indices for this mechanism: $n = m = 2$, $\ell = 1$, $s = 1$, $\delta = 0$.

Example 2 (Lotka 1920):

$$A_1 \xrightarrow{k_1} 2A_1$$

$$A_1 + A_2 \xrightarrow{k_2} 2A_2$$

$$A_2 \xrightarrow{k_3} 0.$$

Mass action kinetics yields the dynamical equations

$$\dot{\rho}_1 = k_1\rho_1(1 - \frac{k_2}{k_1}\rho_2)$$

$$\dot{\rho}_2 = -k_3\rho_2(1 - \frac{k_2}{k_3}\rho_1).$$

The reaction vectors $(1,0)$, $(-1,1)$, $(0,-1)$ span V, so S is the entire species space. There are two stationary points: the origin, and the point $(k_3/k_2, k_1/k_2)$. Linear stability analysis shows the origin to be unstable, and indicates marginal stability for the other stationary point. It turns out that all orbits in V^+ are closed and surround the second point:

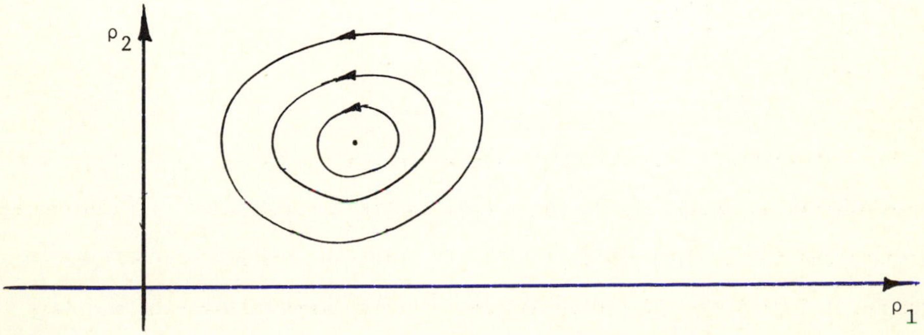

For the Lotka mechanism, $n = 2$, $m = 6$, $\ell = 3$, and $s = 2$, so that $\delta = 6 - 3 - 2 = 1$.

Example 3 (Field and Noyes 1974):

$$H + A_1 \to A_2$$

$$H + A_2 \to 2A_2 + A_3$$

$$A_1 + A_2 \to P$$

$$2A_2 \to P_2$$

$$A_3 \to A_1.$$

This scheme is a simple model of the best known laboratory example of an oscillatory chemical reaction: the Belousov reaction. The model is called the "Oregonator." Although the actual Belousov mechanism involves over 30 different species and is not completely known, the above scheme is a reasonable approximation; in fact, one can identify $A_1 = Br^-$, $A_2 = HBrO_2$, $A_3 = Ce(IV)$, $H = BrO_3^-$. Species A_1, A_2, and A_3 are classed as intermediates, as they are conserved; the net result of the reaction is the conversion of H to P_1 and P_2. But on a shorter time scale, the concentrations of the intermediates oscillate. For our purposes, we consider the concentration of H as large and fixed, and since the P_i are inert, we replace them by 0. This results in the mechanism

$$A_3 \to A_1 \to A_2 \to 2A_2 + A_3$$

$$A_1 + A_2 \to 0 \leftarrow 2A_2.$$

For this scheme, we may calculate $n = s = 3$, $m = 7$, $\ell = 2$, and $\delta = 2$.

There turns out to be a single stationary point in V^+. For certain parameter values it is stable, and for others it is unstable, and there are stable limit cycles.

Note that in the last two examples, those admitting oscillatory solutions, $\delta > 0$. Among chemical reactions, this is apparently the exceptional case, and is the case which may give rise to unusual types of dynamics, such as multiple stationary states and oscillations.

1.9 Colony models

Biological populations are sometimes segregated into homogeneous geographical groups, and even when they are not, models have often been constructed which picture them as being so grouped. For deterministic analyses of migration, selection, and mutation in colony models, see Nagylaki (1977, 1978c). Colony, or "stepping stone," models incorporating stochastic effects have been well studied; I shall not attempt to review the literature here, except to mention Nagylaki's (1978b, 1979) recent results, which are relevant. He has constructed a migration-selection-mutation model with genetic drift accounted for, which becomes a pair of reaction-diffusion equations in the diffusion limit (as described below). His resulting equations are coupled in a nonstandard manner. Sawyer (1977a,b, 1978a,b) has also done a considerable amount of research on questions related to the present discussion. Finally, see the review of stochastic models in Mollison (1976).

We now explore colony models, attaching a general deterministic migration rule. With further assumptions, which we shall try to make explicit, one is again led to a reaction-diffusion system (1.20).

First, consider single-species migration without reaction. Suppose the population is distributed at discrete equally spaced locations on a one-dimensional habitat. (Unequal spacing may also be handled; we leave this generalization to the reader.) Let the locations be $x_j = jh$, where j is any integer, and h is the spacing. Group the population into "species" A_j according to position, and let n_j denote the number of individuals at location j. We consider a reaction mechanism given simply by migration to closest neighboring locations:

$$A_{j-1} \underset{k^+_{j-1}}{\overset{k^-_j}{\leftrightarrows}} A_j \underset{k^-_{j+1}}{\overset{k^+_j}{\rightleftarrows}} A_{j+1} \cdots \quad (\text{all } j), \tag{1.28}$$

and give it mass action kinetics with reaction constants as indicated. Then

$$\frac{dn_j}{dt} = -n_j(k_j^+ + k_j^-) + k_{j+1}^- n_{j+1} + k_{j-1}^+ n_{j-1}. \tag{1.29}$$

We now define

$$\bar{k}_j = \frac{1}{2}[k_j^+ + k_j^-],$$

$$\delta k_j = k_j^+ - k_j^-$$

$$\bar{k}_{j+1/2} = \frac{1}{2}[k_j^+ + k_{j+1}^-],$$

$$\delta k_{j+1/2} = k_j^+ - k_{j+1}^-.$$

The constant $\bar{k}_j$ may be thought of as the average mobility at colony j, whereas $\bar{k}_{j+1/2}$ is the average mobility between colonies j and $j + 1$. The latter would be a property of the terrain between these two colonies. Similarly, δk_j could be termed the degree of directional preference at j, and $\delta k_{j+1/2}$ the degree of directional preference of the migration between colonies j and $j + 1$.

It is natural to speak of the net flux of individuals from j to $j + 1$. It is

$$H_{j+1/2} = k_j^+ n_j - k_{j+1}^- n_{j+1}.$$

From (1.29),

$$\frac{dn_i}{dt} = -H_{j+1/2} + H_{j-1/2}. \tag{1.30}$$

A little manipulation yields two alternative expressions for H:

$$H_{j+1/2} = -\bar{k}_{j+1/2}\Delta n_{j+1/2} + \delta k_{j+1/2} n_{j+1/2}, \tag{1.31}$$

where we have defined $n_{j+1/2} = \frac{1}{2}(n_j + n_{j+1})$, $\Delta n_{j+1/2} = n_{j+1} - n_j$, and

$$H_{j+1/2} = (\bar{k}_{i+1}n_{i+1} - \bar{k}_i n_i) + \frac{1}{2}(\delta k_{i+1}n_{i+1} + \delta k_i n_i). \tag{1.32}$$

Let us now assume the following:

(1) There exist twice continuously differentiable (in x) functions $\rho(x,t)$, $D(x,t)$, $C(x,t)$, with ρ also being continuously differentiable in t, such that (with t fixed)

$$\rho(jh) \equiv \rho_j = h^{-1}n_j,$$

$$D((j + \tfrac{1}{2})h) \equiv D_{j+1/2} = h^2\bar{k}_{j+1/2},$$

$$C((j + \tfrac{1}{2})h) \equiv C_{j+1/2} = -h\delta k_{j+1/2}. \tag{1.33}$$

Then from (1.31),

$$H_{j+1/2} = -D_{j+1/2}\frac{\rho_{j+1} - \rho_j}{h} - C_{j+1/2}\rho_{j+1/2} + C_{j+1/2}(\rho_{j+1/2} - \frac{1}{2}(\rho_j + \rho_{j+1}))$$

$$= -D_{j+1/2}\rho_x((j + \tfrac{1}{2})h,t) - C_{j+1/2}\rho_{j+1/2} + E_j,$$

where $|E_j| \leq Kh$ for some K independent of h, but depending on $\sup|\rho_x|$.

If we now define

$$H = -D\rho_x - C\rho \tag{1.34}$$

we see that (1.30) becomes

$$\frac{\partial\rho}{\partial t} = \frac{\partial}{\partial x}(D\frac{\partial\rho}{\partial x} + C\rho) + O(h).$$

The model now involves neglecting the final term. It apparently works best, then, when the colony size is large (for (1.29) to hold), and h small. If D is a function of moderate size, this latter implies (from (1.33)) that the k_j are large, like h^{-2}. It may be asked, is there any a priori reason why, when h is small, the k_j should behave like h^{-2}? The only suggestive argument is the analogy with Brownian motion, and with the properties of the redistribution kernel k of Section 1.2; these latter properties came out quite naturally.

Corresponding to (1.33), there is an alternative assumption:

(2) There exist twice continuously differentiable (in x) functions ρ, V, and M, with ρ continuously differentiable in t as well, such that (suppressing t)

$$\rho(jh) = h^{-1}n_j,$$

$$V(jh) = 2h^2\bar{k}_j,$$

$$M(jh) = h\delta k_j.$$

Then reasoning in the same manner as before, we are led to the equation

$$\frac{\partial\rho}{\partial t} = \frac{\partial}{\partial x}\left(\frac{1}{2}\frac{\partial}{\partial x}(\rho V) + M\rho\right) + O(h),$$

as in (1.13).

When setting up a model for specific organisms in a specific habitat, it may be simpler to translate the known properties of the organism and habitat in terms of the k_j of the colony model, rather than the redistribution kernel k of Section 1.2. Furthermore, when there is a discontinuity in the habitat, it may be easier to decide upon the proper interface condition this way. Once one is convinced that a RD is appropriate, one could thus set out to construct the specific coefficients and boundary conditions by envisaging a colony situation.

The above treatment can be extended to the case of several species, their interactions producing a source term F, as in (1.20). Both the colony and the continuous model can be extended in a natural way to more space dimensions.

All this was under the hypothesis of an infinite habitat, without boundary. But we can use the same procedure to model boundary effects, if there is a boundary. To see this, let us take (1.28) as holding only for $j < 0$, and picture $x = 0$ as being a boundary. At this boundary, we allow the possibility of individuals being withdrawn from, and/or introduced into, the population (or neither!). We express this by the additional reactions

$$A_o \underset{\hat{k}_1^-}{\overset{\hat{k}_o^+}{\rightleftarrows}} 0. \tag{1.35}$$

Then in addition to (1.29) holding for $j < 0$, we have

$$\frac{dn_o}{dt} = k_{-1}^+ n_{-1} - k_o^- n_o - \hat{k}_o^+ n_o + \hat{k}_1^-. \tag{1.36}$$

Again, let h be small, and for the new reactions set $\hat{k}_o^+ = a_1 h^{-1}$, and $\hat{k}_1^- = a_2$, independent of h. The justification for this is as follows. The mass transfer rate out of the population expressed by the top reaction in (1.35) is $\hat{k}_o^+ n_o = \hat{k}_o^+ h\rho(0)$. We wish this to be a quantity which is independent of h, for given density ρ. Therefore we set $\hat{k}_o^+ h = a_1$. On the other hand, the bottom reaction in (1.35) expresses a mass transfer rate of $\hat{k}_1^-$, which should be independent of h.

Setting $n_j(t) = h\rho(x_j,t)$ in (1.36), we get

$$h\,\frac{\partial\rho(0,t)}{\partial t} = H_{-1/2} - a_1\rho(0,t) + a_2.$$

Using the expression given before for H, and neglecting terms of order h, we obtain (t is suppressed)

$$-D(0)\rho_x(0) - (C(0) + a_1)\rho(0) + a_2 = 0,$$

which is a boundary condition of the third kind at the boundary point $x = 0$.

1.10 Simplifying the model by means of asymptotics

The role of asymptotics in model-building is to reduce complex models to simpler ones, or combinations of simpler ones. This is accomplished by identifying terms in equations, or other influences figuring in the more complex model, which are significantly smaller than others, and neglecting them. The identification here may be enhanced by transforming the equations; typical transformations involve rescaling the variables. In the process of such a reduction, some accuracy and intuitive clarity may be lost, and the range of validity of the model may be restricted.

As far as accuracy is concerned, in population dynamical situations the original (complex) model itself had more than likely been constructed by neglecting a multitude of influences, and so the further reduction we envisage is a continuation of the same process. In many cases (as in the examples described below), a quantitative estimate of the discrepancy between the solutions of the two systems may be established rigorously.

Intuitive clarity may be lost when the original model was constructed from first principles, so that the origin and reason for each term is clear. When the equations are subjected to a mathematical transformation and some terms dropped, they may no longer have such a direct lucid connection with the phenomenon being modeled. A restriction of the conditions under which the model is valid may be needed in order to justify assuming the neglected terms are small.

Among the many rate constants and other parameters figuring in the kinetics for a given reaction mechanism, it is often found that some are of different orders of magnitude than others. This itself may allow an effective reduction in the order of the system of dynamical equations (1.20) or (1.25) through well-known methods of singular perturbations. An example is the "pseudo-steady-state" hypothesis, which is so standard that it is commonly employed in the modeling stage. This approach to the pseudo-steady-state hypothesis was taken first by Bowen, Acrivos, and Oppenheim (1963). See also the accounts in Heineken, Tsuchiya, and Aris (1967), Lin and Segel (1974), and Murray (1977). We shall examine it carefully in this section, in the framework of the kinetic equations

(1.25), and also touch on possible generalizations of the basic method. Examples of the use of asymptotics to simplify (1.20) will be given later, notably in Chapter 2.

Possibly the simplest enzyme reaction mechanism is given by the equation

$$E + S \underset{k_2^*}{\overset{k_1^*}{\rightleftarrows}} C \xrightarrow{k_3^*} E + P. \tag{1.37}$$

In this, the enzyme E catalyzes the conversion of substrate S into the product P. An intermediate complex C is involved in the process. There are four species in this mechanism, and hence four dynamical equations. They can be reduced to two because of the fact (as we shall see) that the dynamics occur on two-dimensional stoichiometric parallels. A further final reduction to a single equation may be made through an approximation, provided a certain smallness condition is met. This final equation involves the well-known Michaelis-Menten rate function. The latter has been used for decades, but only in the 1960's has its mathematical justification been brought to light. In the final reduction. we more or less follow the treatment by Heineken, Tsuchiya, and Aris (1967).

Let e, s, c, and p denote the concentrations of the four species in (1.37). In this four-dimensional space, it is seen that the stoichiometric subspace is spanned by the two vectors $(1,1,-1,0)$ and $(1,0,-1,1)$. Hence $s = 2$. This means that $u = (e,s,c,p)$ must, at all times, be of the form

$$u = u_o + \alpha(1,1,-1,0) + \beta(1,0,-1,1),$$

where $u_o = u(0)$, and α, β depend on t. It follows immediately that

$$e + c = e_o + c_o \tag{1.38}$$

and

$$s + p = e + s_o + p_o - e_o. \tag{1.39}$$

Hence if e and s are known, c and p may be found directly from these equations. Solving the dynamical equations for e and s alone therefore suffices for the determination of the whole orbit. For simplicity, let us assume $c_o = p_o = 0$. Then the differential equations for e and s take the following form, (1.38-1.39) being born in mind:

$$\dot{e} = k_2^* c - k_1^* es + k_3^* c = k_2^*(e_o - e) - k_1^* es + k_3^*(e_o - e), \tag{1.40}$$

$$\dot{s} = k_1^* c - k_1^* es = k_2^*(e_o - e) - k_1^* es. \tag{1.41}$$

This system cannot be solved analytically. But we now assume that e_o is small, and effect an approximation procedure. For the sake of order, we nondimensionalize by setting $e = e_o x$, $s = s_o y$, $\varepsilon = e_o/s_o$, and assume ε to be small. With these replacements, (1.40-1.41) become

$$\dot{x} = (k_2 + k_3) - (k_2 + k_3 + k_1 y)x, \tag{1.42a}$$

$$\dot{y} = \varepsilon[k_2(1 - x) - k_1 xy], \tag{1.42b}$$

where $k_1 = k_1^* s_o$, $k_2 = k_2^*$, $k_3 = k_3^*$, with initial conditions

$$x(0) = 1, \quad y(0) = 1. \tag{1.42c}$$

We first seek a naive approximation x^o, y^o by merely setting $\varepsilon = 0$ in (1.42). So doing, we obtain

$$y^o(t) = \text{const} = 1, \tag{1.43a}$$

$$\dot{x}^o(t) = k - (k + k_1)x^o(t), \quad x^o(0) = 1, \tag{1.43b}$$

where $k = k_2 + k_3$, so that $x^o(t) = \hat{x} + (1 - \hat{x})e^{-(k+k_1)t}$, $\hat{x} = k/(k + k_1)$.

We shall show that this is a legitimate approximation to the solution for ε

small, for any fixed time interval $(0,T)$. However, it will not in general be a good approximation for all time. For example, we may convert (1.42b) into the integral equation

$$y(t) = 1 + \varepsilon \int_0^t [k_2(1 - x(t')) - k_1 x(t')y(t')]dt'.$$

Since $0 \leq x \leq 1$ (this follows from (1.38) with $c_o = 0$) and y, remaining on a stoichiometric parallel, must remain bounded, the integral is some bounded function of t. Therefore

$$|y(t) - 1| \leq \varepsilon M_1 t \tag{1.44}$$

for some M_1 and for all t. Hence our approximation $y^o = 1$ will be valid only for times $t \leq T(\varepsilon)$, where $T(\varepsilon) = o(\varepsilon^{-1})$ as $\varepsilon \to 0$.

Problem: Show that $|x(t) - x^o(t)| \leq \varepsilon M_2 t$.

Hint: Write a differential equation for $x - x^o$, and use (1.44) to show the right side is bounded by $-a(x - x^o) + \varepsilon bt$ for some positive a and b.

Better approximations can be found in the form

$$x(t,\varepsilon) = x^o(t) + \varepsilon x^1(t) + \ldots + \varepsilon^m x^m(t),$$

$$y(t,\varepsilon) = y^o(t) + \varepsilon y^1(t) + \ldots,$$

by substituting these expressions into (1.42) and equating coefficients of powers of ε. But this will still not yield an approximation which is valid for all time.

To obtain one valid for large t, we employ a new, larger time scale by defining $T = \varepsilon t$. Then rewriting (1.42) in terms of this variable, we find

$$\varepsilon x_T \equiv \varepsilon \frac{dx}{dT} = [k - (k + k_1 y)x], \tag{1.45a}$$

$$y_T \equiv \frac{dy}{dT} = k_2 - (k_2 + k_1 y)x. \tag{1.45b}$$

Again, we first seek a naive formal approximation $\bar{x}^o(T)$, $\bar{y}^o(T)$ by writing $\varepsilon = 0$, to obtain

$$\bar{x}^o = \frac{k}{k + k_1\bar{y}^o}, \tag{1.46a}$$

$$\bar{y}^o_T = k_2 - (k_2 + k_1\bar{y}^o)\bar{x}^o = \frac{-k_1k_3\bar{y}^o}{k + k_1\bar{y}^o}. \tag{1.46b}$$

Now (1.46b) may be solved, if we can supply a proper initial value $\bar{y}^o(0)$. Since on the shorter time scale (t), we found that y (but not x) stayed fairly constant, equal to 1, we set

$$\bar{y}^o(0) = 1.$$

The solution of (1.46b) may now be written in closed form. This is left for the reader to do. Knowing it, $\bar{x}^o$ is found from (1.46a). Notice that $\bar{x}^o(0) \neq 1$, so $\bar{x}^o$ cannot be a good approximation near $T = 0$.

As before, better formal approximations may be obtained by setting

$$x(T,\varepsilon) \sim \bar{x}^o(T) + \varepsilon\bar{x}^1(T) + \ldots, \qquad y(T,\varepsilon) = \bar{y}^o(T) + \ldots,$$

substituting into (1.45), and equating coefficients of ε.

We expect $(\bar{x}^o(T),\bar{y}^o(T))$ to be a good approximation to (x,y) for long time, and $(x^o(t),y^o(t))$ to be good for moderate time. That is,

$$(x(t),y(t)) \sim \begin{cases} (x^o(t),y^o(t)) & \text{for} \quad 0 \le t \le \hat{T}(\varepsilon), \\ \bar{x}^o(\varepsilon t),\bar{y}^o(\varepsilon t) & \text{for} \quad \hat{T}(\varepsilon) \le t < \infty, \end{cases}$$

where $\hat{T}(\varepsilon) = o(\varepsilon^{-1})$, $(\hat{T}(\varepsilon))^{-1} = o(1)$ as $\varepsilon \to 0$. This, in fact, can be proved to be the case.

The larger time behavior is usually the important one in biochemical contexts, and so (1.46b) is usually taken to be basic. It describes the decay of substrate with a rate "constant" which is no longer constant, but rather equal to $(-k_1k_3)/(k + k_1y)$, the Michaelis-Menten expression. In practice, the three constants k_1k_3, k, and k_1 are usually determined by fitting this expression to experimental data.

The solution of (1.46) has the property that as $T \to \infty$, $y \to 0$, $x \to 1$. So $e \to e_o$ and, from (1.39), $p \to s_o$. The final effect of the reaction is therefore to convert all of the substrate into the product, and leave as much enzyme as there was originally.

The approximation (1.46) was derived by setting $\varepsilon = 0$ in (1.44). This is tantamount to setting the rate function on the right of (1.45a) equal to zero. Since $e = e_o x$, this function is equal to $\dot{e}$ and, from (1.38) to $-\dot{c}$. Setting this equal to zero is called the "pseudo-steady-state hypothesis." In more complicated biochemical reaction mechanisms, perhaps several intermediates (not substrates nor products, but compounds or complexes involved in the process of converting one to the other), the pss hypothesis is invoked by, again, setting the reaction rates of all the intermediates equal to zero.

Often the rate constants k_i are large. If they are of the order ε^{-1}, then we may write $k_i = \hat{k}_i/\varepsilon$. The above analysis may be repeated using $\hat{k}_i$; the only essential change is that now t itself is on the larger time scale, and $\tau = t/\varepsilon$ the shorter.

The methods of singular perturbations, as in the above example, provide a systematic way to construct approximate solutions in a great variety of problems. Furthermore, they elucidate the smallness assumptions on which the approximation is based, and indicate why the approximation is valid, when it is. Other examples of singular perturbation procedures in chemical kinetics can be found in Schneider, Amundson, and Aris (1972), Murray (1977), and elsewhere. We shall discuss still other aspects of these methods later, but for now give another example, having to do again with the mechanism (1.37).

Let us assume the substrate, as well as the enzyme, concentration is small, so that it is no longer true that $e_o << s_o$; and that the rate constants k_1^* and k_2^*

are correspondingly large. For simplicity, assume these orders of magnitude can be measured by a single small parameter ε, different from the ε used before. We proceed from (1.40), and this time nondimensionalize by setting $e = e_o x$, $s = e_o z$. For x and z, we obtain the equations

$$\dot{x} = (k_2^* + k_3^*) - (k_2^* + k_3^* + k_1^* e_o z)x,$$

$$\dot{z} = k_2^*(1 - x) - k_1^* e_o xz.$$

Assuming $k_1^* e_o$ and k_2^* are large, we set them equal to $\bar{k}_1/\varepsilon$ and $\bar{k}_2/\varepsilon$ respectively, and $k_3^* = \bar{k}_3$. After multiplying by ε, we get

$$\varepsilon\dot{x} = (\bar{k}_2 + \varepsilon\bar{k}_3)(1 - x) - \bar{k}_1 xz, \tag{1.47a}$$

$$\varepsilon\dot{z} = \bar{k}_2(1 - x) - \bar{k}_1 xz. \tag{1.47b}$$

As a first step to the first approximation, we set $\varepsilon = 0$ in (1.47) and replace (x,z) by (x^o,z^o), to obtain the relation

$$\bar{k}_2(1 - x^o) = \bar{k}_1 x^o z^o, \tag{1.48}$$

which simply states that the reversible reaction is balanced. This, however, does not tell us how x and z change with time. For this, we go to the next approximation in (1.47) by putting $x \cong x^o + \varepsilon x^1$, $z \cong z^o + \varepsilon z^1$, and looking at terms of order ε. There results

$$\dot{z}^o = -\bar{k}_2 x^1 - \bar{k}_1 x^1 z^o - \bar{k}_1 x^o z^1 \equiv q,$$

$$\dot{x}^o = q + k_3(1 - x^o) = \dot{z}^o + k_3(1 - x^o). \tag{1.49}$$

But the relation (1.48) between x^o and z^o implies the following relation between the derivatives:

$$-\bar{k}_3\dot{x}^o = \bar{k}_1\dot{x}^o z^o + \bar{k}_1 x^o \dot{z}^o,$$

or

$$\dot{z}^o = -\frac{\bar{k}_2 + \bar{k}_1 z^o}{\bar{k}_1 x^o}\dot{x}^o.$$

So from (1.49)

$$\dot{x}^o = -\frac{\bar{k}_2 + \bar{k}_1 z^o}{\bar{k}_1 x^o}\dot{x}^o + \bar{k}_3(1 - x^o),$$

hence

$$\dot{x}^o = \frac{\bar{k}_3(1 - x^o)}{1 + [(\bar{k}_2 + \bar{k}_1 z^o)/(\bar{k}_1 x^o)]}$$

$$= \frac{\bar{k}_1\bar{k}_3(x^o)^2(1 - x^o)}{x^o(\bar{k}_1 x^o + \bar{k}_2) + \bar{k}_1\bar{k}_2(1 - x^o)}. \qquad (1.50)$$

If an initial value $x^o(0)$ is provided, then this equation may be solved for $x^o(t)$, and y^o thereupon obtained from (1.48). We cannot assume that the proper initial value $x^o(0)$ is 1, nor that $z^o(0) = s_o/e_o$, for (1.48) at $t = 0$ would preclude both of these conditions holding at the same time. The fact is that the above approximation is based on (1.48), which itself cuts down the freedom in initial conditions for x^o and z^o.

This loss of freedom in initial or boundary conditions is usual in singular perturbation procedures, and is remedied by adding x^o and z^o "boundary layer corrections" which are functions of time on a shorter time scale (or, in boundary value problems, functions of space on a shorter space scale) designed to allow the satisfaction of the originally imposed initial conditions, and which vanish as $\varepsilon \to 0$ at fixed positive times.

Accordingly, we next define $\tau = t/\varepsilon$, and consider an approximation of the form

$$x \cong x^o(t) + \tilde{x}(\tau),$$

$$z \cong z^o(t) + \tilde{z}(\tau).$$

To obtain the proper equations for $\tilde{x}$ and $\tilde{z}$, we rewrite (1.47) with the new time variable, then put in the above expressions, at the same time setting $t = \varepsilon\tau$. The result is

$$\tilde{z}_\tau + \varepsilon\dot{z}^o = \bar{k}_2(1 - x^o(\varepsilon\tau) - \tilde{x}(\tau)) - \bar{k}_1(x^o(\varepsilon\tau) + \tilde{x})(z^o(\varepsilon\tau) + \tilde{z}(\tau)),$$

$$\tilde{x}_\tau + \varepsilon\dot{x}^o = \varepsilon k_3(1 - x^o(\varepsilon\tau) - \tilde{x}(\tau)) + \tilde{z}_\tau + \varepsilon\dot{z}^o.$$

For the lowest order approximation, we again set $\varepsilon = 0$ to obtain

$$\tilde{z}_\tau = \bar{k}_2(1 - x^o(0) - \tilde{x}) - \bar{k}_1(x^o(0) + \tilde{x})(z^o(0) + \tilde{z}), \tag{1.51a}$$

$$\tilde{x}_\tau = \tilde{z}_\tau. \tag{1.51b}$$

The initial conditions for $\tilde{x}$ and $\tilde{z}$ must be adjusted so that $x(0) = 1$, $z(0) = s_o/e_o$, hence

$$\tilde{x}(0) = 1 - x^o(0), \quad \tilde{z}(0) = \frac{s_o}{e_o} - z^o(0). \tag{1.52}$$

But (1.51b) implies $\tilde{x}(\tau) - \tilde{x}(0) = \tilde{z}(\tau) - \tilde{z}(0)$, hence

$$\tilde{x}(\tau) - \tilde{z}(\tau) = (1 - x^o(0)) - (s_o/e_o) + z^o(0). \tag{1.53}$$

As mentioned before, to be boundary layer corrections, $\tilde{x}$ and $\tilde{z}$ must vanish as

$\tau \to \infty$. Passing to the limit in (1.53), we obtain the relation

$$x^o(0) - z^o(0) = 1 - \frac{s_o}{e_o} .$$

This, together with (1.48) at $t = 0$, tells us

$$x^o(0) = \frac{\bar{k}_2(1 - x^o(0))}{\bar{k}_1 x^o(0)} + 1 - \frac{s_o}{e_o} .$$

Solving for $x^o(0)$, we have

$$x^o(0) = \frac{1}{2\bar{k}_1}\left[(1 - \frac{s_o}{e_o} - \bar{k}_2) \pm \sqrt{(1 - \frac{s_o}{e_o} - \bar{k}_2)^2 + 4\bar{k}_1\bar{k}_2}\right].$$

One of these roots is positive and the other negative. We take the positive one, and use it in conjunction with (1.50) to find $x^o(t)$. Then $z^o(t)$ is found from (1.48). We next substitute the expression (1.53) for $\tilde{x}(\tau)$ into the right side of (1.51a), and solve the resulting differential equation for $\tilde{z}$, of course imposing the initial conditions (1.52). It may be checked that the solution $\tilde{z}$ does indeed decay to 0 automatically as $\tau \to \infty$. After this, $\tilde{x}(\tau)$ is obtained from (1.53).

The four functions being thus determined, we finally write our approximate solution as

$$x(t) \cong x^o(t) + \tilde{x}(t/\varepsilon), \tag{1.54a}$$

$$z(t) \cong z^o(t) + \tilde{z}(t/\varepsilon). \tag{1.54b}$$

This, of course, is only a lowest order approximation; higher order terms may be obtained in a systematic manner.

It may now be asked, are these two time scales sufficient for the complete analysis, or do we need another very large one? It may be verified from (1.50) that $x^o(t) \to 1$ as $t \to \infty$, and therefore, from (1.48) that $z^o(t) \to 0$. Therefore the approximations we have found behave in a regular manner as $t \to \infty$ and do what one would expect them to do. One's intuition, then, would say that the approximation (1.54) is uniformly valid for all time, and no further time scale is needed. This can be proved by an argument along the lines of Hoppensteadt (1966).

2. FISHER'S NONLINEAR DIFFUSION EQUATION AND SELECTION-MIGRATION MODELS

A large number of studies have been made of the statics and dynamics of gene frequencies in a population subject to selection pressures and whose individuals may migrate geographically. In the case of deterministic models with space and time continuous, and selection restricted to a single locus with two alleles, the problem may sometimes be reduced to a single nonlinear diffusion equation. Such a reduction is advantageous, as it permits many qualitative properties of the frequency distribution to be obtained without a great deal of difficulty. This is especially true in the investigation of clines when the geographic variation in the selection characteristics is of a simple nature. It should also be brought out that the relevance of nonlinear diffusion equations in population genetics has provided part of the motivation behind the considerable mathematical work which has been devoted to these equations in recent years. This chapter is concerned with the justification for the use of single diffusion equations in modeling selection-migration phenomena. We begin with an overview of arguments which have been used in the past to model the problem by a single equation. We shall then approach the problem from the point of view of the previous chapter, considering selection to occur at a single gene locus and writing the corresponding reaction-diffusion equations for the three genotypes (which will be our "species"). Asymptotic methods, operating under explicit assumptions, will be used to reduce the three equations to a single one. Throughout the chapter we restrict attention, for simplicity, to populations in a one-dimensional habitat.

2.1 Historical overview

The simplest case of the nonlinear diffusion equation in question is

$$p_t = p_{xx} + sp(1 - p), \qquad (2.1)$$

where p is the frequency of an advantageous gene, and s a measure of intensity of selection. This equation was first written down by Fisher (1937). It has long been known that under various sorts of simplifying assumptions (Fisher 1930, Wright 1969 Chapter 2, Kimura and Crow 1969) the rate of increase in frequency p of a favorable gene in a population is given by the expression $sp(1 - p)$ for some constant s. (In diploid populations, one of these assumptions is that Hardy-Weinberg gene frequency proportions hold at all times.) Fisher simply added to this expression a diffusion term p_{xx}. As justification, Fisher assumed a type of Fick's law to hold, which in this case says that the spatial "flux" of p should be proportional to $-p_x$ if the individuals migrate at random (thus p flows from regions of high to low densities). This assumption is more reasonable for fluxes of individuals than for frequencies; but if the total population density is constant (another assumption), then the flux of individuals is proportional to the flux of p, and Fick's law gains some credibility.

Kolmogorov, Petrovskiĭ, and Piscunov (1937) also studied an equation similar to (2.1), and in justifying the diffusion term, assumed the existence of a function $k(r)$ such that $k(r)dr$ is the probability of an individual migrating a distance in the range $(r, r + dr)$ in unit time (one generation). Then, again under the important assumption that the total population density is constant, the change in frequency Δp from one generation to the next due to migration alone is given by

$$\Delta p(x,t) = \int_{-\infty}^{\infty} k(x - y)p(y,t)dy - p(x,t).$$

Now suppose k is even, has small but positive second moment ℓ^2 with higher moments small relative to ℓ^2, suppose p is smooth in x and t, and suppose s and ℓ are small. Then approximately

$$\Delta p(x,t) = \frac{\ell^2}{2} p_{xx}(x,t).$$

Combining this change due to migration with a change due to selection based on Fisher's reasoning, the authors obtained a nonlinear diffusion equation similar to (2.1).

Haldane (1948) employed a space-dependent version of (2.1),

$$p_{xx} + s(x)p^2(1 - p) = 0 \tag{2.2}$$

in initiating the analytical study of clines (spatially-varying time-independent distributions) in a continuous habitat. Here he took $s(x)$ to be discontinuous at $x = 0$, to change sign there, and to be constant otherwise. Thus, the direction of selection changes abruptly at $x = 0$. The diffusion term was justified by an argument similar to that of Kolmogorov et al. Also as in the latter paper, dominance is assumed; this results in the factor p^2 in (2.2) as distinct from (2.1). Haldane made his biological assumptions more explicit than did the previous authors; besides those indicated above, he mentioned constancy of the density, random mating, synchronization of generations, and separate seasons for migrating and breeding. Whereas Fisher (1937) and Kolmogorov et al. (1937) were concerned with traveling population waves, Haldane's interest was in stationary clines.

Returning to his selection-migration model, Fisher (1950) posited linear spatial dependence of the selection parameters, and investigated the resulting cline solution of

$$p_{xx} + xsp(1 - p) = 0. \tag{2.3}$$

This represents a smooth variation in environmental factors, as opposed to Haldane's abrupt change. Fisher provided a numerical solution of (2.3), and a discussion of scaling considerations.

Montroll and West (1973) showed how a variety of simplistic random-walk migration models can be reduced to a single nonlinear diffusion equation. In this connection, see also Skellam (1951).

Slatkin (1973) continued the work of Haldane and Fisher by studying the equation

$$\frac{\ell^2}{2} p_{xx} + s\gamma(x)p(1 - p) = 0 \tag{2.4}$$

where ℓ, again, is a typical dispersion length (in fact ℓ^2 is the same second moment used by Kolmogorov et al. and Haldane). It is plausible from (2.4), and Slatkin brought out, that the characteristic length of spatial variation in a cline is $\ell_c = \ell/\sqrt{s}$, provided the characteristic length of environmental variation (in $\gamma(x)$) is no greater than that. Slatkin's assumptions are similar to Haldane's, but he gave a different derivation of the diffusion term. Slatkin generalized (2.4) to cases of various degrees of dominance; of particular interest is the case of greatly reduced heterozygote fitness, for which cline steepness is somewhat independent of the steepness of the environmental gradient. He also modeled geographical barriers.

Using the same model, May, Endler, and McMurtrie (1975) further discussed, among other things, questions relating to characteristic length scales.

Nagylaki (1975) generalized the model by allowing the dispersal and selection characteristics to depend on space and time, and allowing regulation of the population size by a logistic type density dependence in the selection term. As usual, it was assumed that Hardy-Weinberg proportions hold locally. In his derivation of the model nonlinear diffusion equation, Nagylaki assumed migration, breeding, and regulation with selection to occur separately. He accounted for migration with a dispersal function $k(x,t;y,t+\Delta t)$ which may be heterogeneous; i.e., it is not just a function of $x - y$ and Δt, as previous authors had assumed. Assumptions were made about the limiting behavior as $\Delta t \to 0$ of certain integrals involving k; this in a sense replaces Kolmogorov et al.'s assumption about the smallness of ℓ, for example. On this basis, Nagylaki gave a more satisfactory derivation of the diffusion approximation.

But we need to say what it means for $\Delta t \to 0$. Two conceptual possibilities occur here. If the selection parameter s is small and generations are synchronized, one could argue that time should be measured in units of $1/s$ generations, as this is how long it takes for any change to occur. Then one generation occurs in time $\Delta t = s$, and small s means small Δt. Alternatively, one could suppose that breeding, etc., is not synchronized, and that throughout the large population these activities occur simultaneously. Then time is naturally continuous, and we may let $\Delta t \to 0$. In this case, however, the assumption of Hardy-Weinberg proportions is not so clear.

The equation that Nagylaki obtains at this point is difficult to handle, but becomes more tractable under the additional assumption that the population has constant density (its carrying capacity). (The cases Nagylaki actually analyzes also have constant migration rate.)

In the case of one space dimension, the term accounting for migration is as on the right side of (1.13).

In later papers (1976, 1978a,b, 1979), Nagylaki considered (a) a special case of spatial dependence in V: a step function with one discontinuity (1976); (b) the effects of drift (1978b, 1979); (c) a model in which the population density undergoes a discontinuity at one point (1978a); and (d) the effects of a geographical barrier (1976). In (a) and (d), a discrete migration model was used to deduce the appropriate interface condition at the discontinuity. Sawyer (1978a) has made a thorough analysis of these and related issues.

Aronson and Weinberger (1975) used a different approach. Not assuming Hardy-Weinberg proportions or constant density at the outset, they began (as shall we) with separate diffusion equations for the separate genotypes. They assumed constant birth and death rates, therefore density- and frequency-independent selection, and no population regulation. From any solution of this system, the frequency $p(x,t)$ of one of the genes is easily determined. Aronson and Weinberger asked whether this function can be approximated by an appropriate solution of a scalar nonlinear diffusion equation analogous to (2.1). They chose the solution of the latter which agrees with the given p at some initial instant of time $t = 0$. Then if the selection is weak, the birth and death rates large, and the initial spatial gradient of the solution is small, they found good agreement up to times of order $O(s^{-1})$, where s is a measure of the intensity of selection. In the main part of the paper, they proceeded to prove results about Fisher-type equations with no x-dependence, relative to stability and wave front propagation.

Further mathematical analysis relating to clines for equations generalizing (2.1) and (2.4) was done by Conley (1975), Fleming (1975), Fife and Peletier

(1976), and Peletier (1976). Much analysis on front propagation has also been done; see the bibliography in Fife and McLeod (1977); also see Conley and Fife (1979).

In the derivations outlined above, excepting that of Aronson and Weinberger, pervasive use has been made of three basic assumptions (among others): (1) the population maintains Hardy-Weinberg proportions; (2) the density is constant; and (3) the selection mechanism is of a rather simple type, involving constant relative viability of the three genotypes.

Assumption (1) is justified if the generations are synchronized and mating is truly random, but a more precise argument needs to be put forth in cases when these conditions do not necessarily hold. Assumption (2) may be approximately true when there is little environmental diversity in space or time and the selection is weak; for then the carrying capacity is approximately independent of gene frequencies. But when the selection pressure depends moderately or strongly on x and/or t, density cannot be expected to be constant.

2.2 Assumptions for the present model

The goal now is to examine the validity of scalar nonlinear diffusion equation models for the migration-selection problem at hand, when selection is weak, ridding ourselves of assumptions (1) and (2) mentioned above, and allowing a greater degree of generality in the selection process (3). Our starting point is a system of reaction-diffusion equations for the genotype frequencies, generalizing those with which Aronson and Weinberger (1975) began their analysis. We then proceed by a systematic formal asymptotic analysis based on the smallness of a selection parameter. In this regard our approach is similar to Hoppensteadt's (1976), who did not consider migration. For an analysis which is more than formal, see Conley and Fife (1979).

In our model, certain assumptions made in the papers cited in Section 2.1 will be retained:

1) The population is large enough so that

   (a) genetic drift and other stochastic effects may be neglected,

   (b) the densities of the genotypes are well-defined functions of x and t, these being continuous variables,

(c) their birth and death rates are well-defined functions of x, t, and the densities of all genotypes.

2) Selection occurs at one gene locus, with two available alleles (or groups of alleles).

3) The age and sex structure of the population does not affect its birth, death, or migration rates.

For the migration effects, we operate in the framework of the assumptions in Chapter 1. This leads to a reaction-diffusion system of type (1.20). Alternatively, under stated additional assumptions about the redistribution kernel, we obtain (1.20) with the spatial derivative parts replaced by the right side of (1.13). Since it is more traditional to use the infinitesimal variance V and drift M in population genetics (Nagylaki 1975), we take the latter approach. The source function F in (1.20) can be written as follows in terms of the birth rates r_b^i and death rates r_d^i (the index i specifies the genotype):

$$F_i(\underline{\rho}) = \rho_i(r_b^i(\underline{\rho}) - r_d^i(\underline{\rho})).$$

In all, there results a system

$$\partial_t \rho_i = \partial_x\left(\frac{1}{2}\, \partial_x(V\rho_i) + M\rho_i\right) + \rho_i(r_b^i - r_d^i), \tag{2.5}$$

where V, M, r_d, r_b depend on x and t, and r_d, r_b depend on $\underline{\rho}$ as well. One could also allow V and M to depend weakly, in a certain sense, on $\underline{\rho}$, but we shall not pursue this.

Our point of departure will be (2.5), together with a specification of the form of the selection processes (which amounts to a specification of how the r_b^i, r_d^i may be allowed to depend on the ρ_i). The selection, though weak, may act through mating preferences, fertility rates, and death rates. They may be conditioned by such things as crowding and competition, so that selection is density-dependent. For another approach to modeling the rate functions, see Nagylaki and Crow (1974). All this complication is reduced at the end to a rather simple nonlinearity in the Fisher equation which we derive.

A final assumption we make is that a carrying capacity for the population exists, which depends only slightly on gene frequency, but may depend strongly on x and t. (The slight dependence on gene frequency is not necessary; we get a similar nonlinear diffusion equation at the end when the dependence is strong, but strong dependence is unrealistic in view of the weakness of the selection.)

Though we do not pursue this question, our analysis extends immediately to models in which more than two alleles are available for the locus in question. If m is the number of such alleles, one obtains at the end a system of $m - 1$ nonlinear diffusion equations for the gene frequencies. For $m > 2$, any analysis of this system would be an order of magnitude more difficult than it would for the scalar case.

First, we describe the form our model takes when no selection is present. We imagine a large single-species population of randomly migrating individuals whose density ρ is continuous in space and time. Space is one-dimensional; this restriction is for convenience only, and the argument can be readily extended to two-dimensional problems. Rather than speaking of fitness or Malthusian parameters, we find it convenient to distinguish birth and death rates r_b and r_d. Our basic dynamic equation for ρ is

$$\partial_t\rho - \partial_x\left(\frac{1}{2} \partial_x(V\rho) - M\rho\right) = \rho(r_b - r_d). \tag{2.6}$$

Here r_b and r_d are allowed to depend on x, t, and ρ.

We now put an added complication into the model, by supposing the population to be diploid and distinguishing between genotypes according to genetic content of one locus. We suppose only two alleles A_1, A_2 to be available. The genotypes are denoted by A_iA_k $(i,k = 1, 2)$, and their densities by ρ_{ik}, so that $\sum_{i,k} \rho_{ik} = \rho$. (Genotypes A_iA_k and A_kA_i are really the same; but we distinguish them for notational convenience, using $\rho_{ik} = \rho_{ki}$. In this regard we follow the notation of Crow and Kimura (1970).) Let $p_i = \left(\sum_k \rho_{ik} \right)/\rho$ denote the frequency of allele A_i in the population.

Still assuming no selective difference between the genotypes, if completely random mating occurs and the densities of males and females are identical, then the fraction of genotype A_iA_k among the total births will be equal to p_ip_k. Similarly the fraction of genotype A_iA_k among the total deaths will be equal to the same fraction in the whole population, that is $p_{ik} \equiv \rho_{ik}/\rho$.

Equation (2.6) is now replaced by a system for the various density functions ρ_{ik}. According to the above, the rate of production of ρ_{ik} is $\rho r_b p_i p_k - \rho r_d p_{ik}$:

$$\partial_t \rho_{ik} - \partial_x \left(\frac{1}{2} \partial_x (V\rho_{ik}) - M\rho_{ik} \right) = \rho (r_b p_i p_k - r_d p_{ik}). \qquad (2.7)$$

Our next assumption is that there is a carrying capacity $\rho_c(x,t)$ for each (x,t). This means that

$$r_b(x,t,\rho) - r_d(x,t,\rho) \begin{cases} > 0 & \text{for} \quad \rho < \rho_c(x,t) \\ = 0 & \text{for} \quad \rho = \rho_c(x,t) \\ < 0 & \text{for} \quad \rho > \rho_c(x,t). \end{cases} \qquad (2.8)$$

At this point we insert a selection mechanism into the model by varying the birth and death rates in (2.7) by a small amount. Let f_{ik}, a function of x, t, and all the $\rho_{j\ell}$, be the fraction of genotypes A_iA_k among the total births. Since no individual of this type can be born when there are no alleles A_i in the population, and the same holds when there are no A_k, we have that $f_{ik} = 0$ when either $p_i = 0$ or $p_k = 0$. Therefore we may write $f_{ik} = p_i p_k g_{ik}$. As we have seen, under no selection, $g_{ik} = 1$; we now perturb it by setting $g_{ik} = 1 + s\eta_{ik}$ for some function η_{ik} of x, t, and all the $\rho_{j\ell}$. Here s is a small parameter. By the analogous argument, we replace $r_d p_{ik}$ in (2.7) by $r_d p_{ik}(1 + s\gamma_{ik})$. The result is

$$\partial_t \rho_{ik} - \partial_x \left(\frac{1}{2} \partial_x (V\rho_{ik}) - M\rho_{ik} \right) = \rho [r_b p_i p_k (1 + s\eta_{ik}) - r_d p_{ik}(1 + s\gamma_{ik})]. \qquad (2.9)$$

2.3 Reduction to a simpler model

Summing on i and k, and using $\Sigma p_i p_k = \Sigma p_{ik} = 1$, one obtains from (2.9)

$$\partial_t \rho - \partial_x\left(\frac{1}{2}(V\rho)_x - M\rho\right) = \rho[r_b - r_d + sr_b \Sigma p_i p_k \eta_{ik} - sr_d \Sigma p_{ik}\gamma_{ik}]. \quad (2.10)$$

The dynamics of p_{ik} can be found from (2.9) and (2.10):

$$\partial_t p_{ik} = \frac{1}{\rho}\partial_t \rho_{ik} - \frac{p_{ik}}{\rho}\partial_t \rho = \frac{1}{2\rho^2 V}\partial_x(V^2\rho^2\partial_x p_{ik}) - M\partial_x p_{ik} + r_b(p_i p_k - p_{ik}) +$$

$$+ s\{r_b(p_i p_k \eta_{ik} - p_{ik}\bar{\eta}) - r_d p_{ik}(\gamma_{ik} - \bar{\gamma})\}, \quad (2.11)$$

where $\bar{\eta} = \sum_{k,\ell} p_k p_\ell \eta_{k\ell}$, $\bar{\gamma} = \sum_{k,\ell} p_{k\ell}\gamma_{k\ell}$.

Summing over k, we find

$$\partial_t p_i - \frac{1}{2\rho^2 V}\partial_x(V^2\rho^2\partial_x p_i) - M\partial_x p_i = s\{r_b p_i(\bar{\eta}_i - \bar{\eta}) - r_d(\sum_k p_{ik}\gamma_{ik} - p_i\bar{\gamma})\}, \quad (2.12)$$

where $\bar{\eta}_i = \sum_k p_k \eta_{ik}$ ($\bar{\gamma}_i$, defined analogously, will be used later).

We wish to exploit the smallness of s to obtain a simplification of these equations which is both meaningful and tractable. One approach to suggest itself would simply be to set $s = 0$ in them. But this naive procedure amounts to ignoring selection altogether, and would therefore not give us any new information. Instead, to account for the effects of weak selection, one must recognize that they are seen only after a sufficient amount of time has elapsed. This suggests the use of a long time scale, with unit related to the parameter s. This is plausible from (2.12); in order to observe any temporal and spatial variation in p, it is necessary to adjust the time and length scales so that the terms on the left are of the same order of magnitude as those on the right.

While we are doing this, it is appropriate to make everything nondimensional. Accordingly, we let r_o and V_o be a typical rate and a typical migration rate (such as the maxima of r_b and V) and define

$$\tau = sr_o t, \quad \xi = \sqrt{\frac{sr_o}{V_o}}\, x, \quad \alpha_{b,d} = \frac{r_{b,d}}{r_o}, \quad \Delta = \frac{V}{V_o}, \quad \mu = \frac{M}{\sqrt{sr_o V_o}} .$$

Then (2.12) with $i = 1$, (2.11) with $(i,j) = (1,2)$, and (2.10) become the following, where we have defined $p = p_1$:

$$\partial_\tau p - \frac{1}{2\rho^2\Delta} \partial_\xi(\rho^2\Delta^2\partial_\xi p) + \mu\partial_\xi p = h_1, \tag{2.13a}$$

$$s(\partial_\tau p_{12} - \frac{1}{2\rho^2\Delta} \partial_\xi(\rho^2\Delta^2\partial_\xi p_{12} + \mu\partial_\xi p_{12})) = \alpha_b(p_1p_2 - p_{12}) + sh_2, \tag{2.13b}$$

$$s(\partial_\tau \rho - \frac{1}{2} \partial_\xi^2(\Delta\rho) + \partial_\xi(\mu\rho)) = \rho(\alpha_b - \alpha_d) + sh_3, \tag{2.13c}$$

where h_i are specific functions of x, t, p_{ik}, and ρ.

In the three equations (2.13) (which have now replaced our original system (2.9) as basic equations), we do obtain a meaningful first approximation by setting $s = 0$. Doing this amounts to assuming that the terms so discarded are small in some sense. This question will be discussed more fully in Sections 2.4 and 2.5. At this point, suffice it to be said, rather imprecisely, that if the given data of the problem $(\alpha_{d,b}, \Delta, \eta_{ij}, \gamma_{ij})$ are smooth and have spatio-temporal variation on scales no smaller than those of ξ and τ, and our concern is with steady-state solutions, then dropping the terms in (2.13) with small coefficient s is formally justified. It is also sometimes justified when these conditions are not met (as for the example in Section 2.6), but caution should be exercised.

We therefore proceed to set $s = 0$ in (2.13). From (2.13b), we obtain

$$p_{12} = p_1p_2, \tag{2.14}$$

and from (2.13c), $\alpha_b = \alpha_d$, which implies

$$\rho = \rho_c . \tag{2.15}$$

The relation (2.14) says that, to first approximation, Hardy-Weinberg proportions hold, and (2.15), that we are at carrying capacity. Substituting (2.14) and (2.15) into (2.13a), we obtain

$$\partial_\tau p - \frac{1}{2\rho_c^2\Delta} \partial_\xi (\rho_c^2 \Delta^2 \partial_\xi p) + \mu \partial_\xi p = f(x,t,p), \tag{2.16}$$

where

$$f = \alpha_b p(\bar{\eta}_1 - \bar{\eta}) - \alpha_d(\bar{\gamma}_1 - \bar{\bar{\gamma}})$$

$$= \alpha_b p[(\bar{\eta}_1 - \bar{\eta}) - (\bar{\gamma}_1 - \bar{\bar{\gamma}})]$$

$$= \alpha_b p(1 - p)[(\bar{\eta}_1 - \bar{\eta}_2) - (\bar{\gamma}_1 - \bar{\gamma}_2)].$$

Setting $\omega_{11} = \eta_{11} - \gamma_{11}$, $\omega_{12} = \eta_{12} - \gamma_{12}$, $\omega_{22} = \eta_{22} - \gamma_{22}$, the right sides all being evaluated at Hardy-Weinberg equilibrium and carrying capacity, we obtain

$$f = \alpha_b(x,t)p(1 - p)g(x,t,p),$$

where

$$g = p(\omega_{11} - \omega_{12}) + (1 - p)(\omega_{12} - \omega_{22}),$$

and $\alpha_b(x,t)$ is simply α_b at carrying capacity.

Fisher's (1937) original equation (2.1) is obtained from (2.16) in the event that none of the given data depend on x or t, the ω_i are constant, and $\omega_2 = \frac{1}{2}(\omega_1 + \omega_3)$. The other particular cases studied in the past are also obtained by setting the ω_i equal to various constants.

2.4 Comments on the comparison of models

The selection-migration model (2.16) was obtained via the following steps: (a) We gave explicit biological assumptions, which led to a mathematical model in the form of a reaction-diffusion system (2.9); (b) by rescaling and rearranging terms, we derived the equivalent set of equations (2.13); (c) we discarded terms in (2.13) which are, formally, of small magnitude, to obtain the single equation (2.16), a much simpler model to work with than (2.9).

The original model (2.9) is intuitively justifiable, as the significance of each term in those equations can be understood. The final model (2.16) is not so intuitively obvious. Nevertheless it was obtained from a justifiable model by neglecting terms which, one can argue, are small in commonly occurring situations. Since neglecting influences which are small is the essence of model building, in those situations (2.16) stands on its own right as a legitimate model for the biological picture originally envisaged. At the same time, however, it is satisfying to be able to lend rigor to the comparison of two models, such as (2.9) and (2.16).

Therefore two general questions arise at this point, regarding the legitimacy of (2.16) as a model:

1) Are the terms of (2.13) neglected in passing to (2.16) indeed small, and if so, in what sense are they? This is essentially a question about the formal justification for (2.16).

2) Can the two models (2.13) and (2.16) be compared in a more rigorous manner?

We shall take up the first of these questions in Section 2.5. Sufficient conditions for the correctness of (2.16), as a formal approximation to (2.13), will be discussed, and it will be shown in Section 2.6 that even in some cases when the discarded terms are not uniformly small, the approximation is still justified.

As for the second question, it should be noted that lending rigor can be interpreted different ways, but in any case it should mean proving that the important properties of one model are approximately true of the other as well. In the present case, the properties of nonlinear diffusion equations such as (2.16) which have been of most interest are those relating to clines and wave fronts. It would therefore be of mathematical interest to prove that (i) the existence of a stable cline solution of (2.16) implies the existence of a cline solution of (2.9) which is close to the former in some sense; (ii) the existence of a wave front for (2.16) implies one for (2.9) which, again, is close; and (iii) every stationary cline or wave front for (2.9) can be discovered this way.

Regarding (ii), if $\alpha_{b,d}$, Δ, η_{ij}, γ_{ij} are independent of (x,t), $f_p(0) < 0$ $f_p(1) < 0$, and a wave front exists for (2.16) (this will be true under nonrestrictive assumptions), then Conley and Fife (1979) have established the existence of a wave front solution of (2.9) which approximates, in some sense, the former. Aronson and Weinberger (1975) have taken a different approach, comparing the solutions of comparable initial value problems for their analogs of (2.9) and (2.16). Other than these, rigorous results of this type remain to be established.

2.5 The question of formal approximation

Let us return to question the legitimacy of discarding the terms in (2.13) involving s. If these terms are indeed "small" for the solution of interest to us, then we say that omitting them is formally justified. (We shall show by example, however, that the formal reduction of (2.13) to (2.16) may be justifiable in other cases as well.) To be more precise, let $p(\xi,\tau)$ be a solution of (2.16) of interest (in most cases, it will be a stationary solution (cline) or a traveling front). With this solution, we define $\tilde{p}_{12} = p(1 - p)$ and $\tilde{N} = N_c(x,t)$, and ask whether the quantities multiplied by s in (2.13b) are subject to some reasonable bound. If so, then our solution of the simple model (2.16) will be a formally approximate solution of (2.13) as well. What constitutes a reasonable bound is a matter of taste, and to this extent the definition of formal approximation will also be ambiguous. An alternative approach would be to speak in terms of families of problems and solutions, depending on the parameter s, which approaches 0. Then a definition of formal approximation could be given which is mathematically more satisfying, but of less value in practice.

Now suppose the functions f, ρ_c, Δ, and μ in (2.16) vary smoothly in the scales of ξ and τ, in the sense that their first few derivatives with respect to ξ and τ are bounded uniformly by some number K. Then standard a priori estimates for solutions of parabolic equations (Friedman 1964) show that for large

τ, every solution p of (2.16) will, together with its derivatives $p_{\xi\xi}$ and p_τ, be bounded uniformly by some constant $F_1(K)$, where the function F_1 is independent of s. The same is then true of $p_{12} = p(1 - p)$, and the terms in (2.13) involving s will be bounded uniformly by $sF_2(K)$. The magnitudes of K and s, then, determine the degree to which our approximation is formally justified in the above sense.

When the above supposition does not apply, caution has to be exercised in attempting to make the reduction to (2.16).

2.6 The case of a discontinuous carrying capacity

This is an example in which the arguments of Section 2.5 are inapplicable, and caution is necessary. Suppose that all the given data of the problem vary smoothly in space with the length scale of ξ or longer, and are independent of t, except that $\rho_c(x)$ has a discontinuity at $x = 0$. Thus ρ_c jumps from $\rho^- = \rho_c(0-)$ to $\rho^+ = \rho_c(0^+)$. This means that r_b and/or r_d are also discontinuous at $x = 0$.

Our first question is whether (2.9), whose right side is discontinuous at one point, is still a valid model. The arguments used in Sections 1.1 and 1.2 to obtain the form of the terms on the left of (2.9) supposed ρ to be twice differentiable in x, which it won't be at $x = 0$, because of the stated discontinuity. But (see the similar argument in Section 1.3) (2.9) remains valid for $x \neq 0$. What are needed are interface conditions to match the solution of (2.9) for $x < 0$ with the solution for $x > 0$. These are easily provided by considering the phenomenon to be modeled. On the one hand, the densities ρ_{ik} should be continuous at $x = 0$, otherwise random migration would quickly smooth them out. Secondly, the flux $-\frac{1}{2}\partial_x(V\rho_{ik}) + M\rho_{ik}$ should be continuous, to prevent a build-up of individuals at the one point $x = 0$. But solutions with these continuity properties are exactly what one traditionally means by "solutions" of a parabolic differential equation with discontinuous right side. So we can say that in the present circumstances, (2.9) continues to be valid.

The next question is about the validity of our subsequent manipulations of (2.9). Nothing is controversial about the passage to (2.13). But it happens that the terms in (2.13) formally of order s are no longer uniformly small. Nevertheless, (2.16) is still the appropriate reduced problem.

Considering (2.13c), we see that $\rho = \rho_c(x)$ is not an approximate solution, because $\partial_\xi \rho$ will not exist at $x = 0$. For this reason, it is appropriate to examine ρ near $x = 0$ by using a smaller length scale. Arguing physically, we can say that the action of diffusion will smooth out the discontinuous approximation $\rho = \rho_c$, but that the smoothing occurs only in a neighborhood of the origin, on a scale with unit of the order $\sqrt{V_o/r_o}$ (measured in x). So to construct an approximation near $x = \xi = 0$, we rewrite (2.13c) in terms of the variable

$$y = \xi/\sqrt{s} = \sqrt{r_o/V_o}\, x,$$

dropping the time dependence and, as before, neglecting the term sh_3, Let $\hat{\rho}(y)$ be the density ρ, measured in this smaller length scale. Then

$$\frac{1}{2}\partial_y(\Delta(\sqrt{s}y)\partial_y\hat{\rho}) + \hat{\rho}(\alpha_b(\sqrt{s}y,\hat{\rho}) - \alpha_d(\sqrt{s}y,\hat{\rho})) = 0.$$

Restricting attention first to the side $y > 0$, we may set $s = 0$ in this equation, because Δ, α_b, and α_d depend smoothly on $\xi = \sqrt{s}y$. We do the same for the side of $y < 0$. This results in

$$\frac{1}{2}\partial_y^2\hat{\rho} + G(\hat{\rho}) = 0, \tag{2.17}$$

where

$$G(\hat{\rho}) = \begin{cases} \dfrac{1}{\Delta(0)}\hat{\rho}[\alpha_b(0+,\hat{\rho}) - \alpha_d(0+,\hat{\rho})] \equiv G^+(\hat{\rho}), & y > 0 \\ \dfrac{1}{\Delta(0)}\hat{\rho}[\alpha_b(0-,\hat{\rho}) - \alpha_d(0-,\hat{\rho})] \equiv G^-(\hat{\rho}), & y < 0. \end{cases}$$

The solution of this equation would be the basic approximation of the "inner expansion" in a "transition layer" at the origin (Fife 1976a). To match it with the "outer" solution $\hat{\rho} = \rho_c$ on the right and left, we impose the boundary conditions

$$\hat{\rho}(\pm\infty) = \rho^{\pm}. \tag{2.18}$$

This is a standard technique in matched asymptotic analysis.

We shall show that the transition layer problem (2.17), (2.18) does have a solution.

Suppose, for definiteness, that $\rho^+ > \rho^-$. By definition of carrying capacity,

$$\alpha_b(0\pm,\hat{\rho}) - \alpha_d(0\pm,\hat{\rho}) \begin{cases} > 0 & \text{for} \quad \hat{\rho} < \rho^{\pm} \\ < 0 & \text{for} \quad \hat{\rho} > \rho^{\pm}. \end{cases}$$

Therefore for $\rho^- < \hat{\rho} < \rho^+$,

$$G(\hat{\rho}) \begin{cases} > 0 & \text{for} \quad y > 0 \\ < 0 & \text{for} \quad y < 0. \end{cases}$$

The maximum principle then implies that any solution of (2.17), (2.18) must in fact satisfy $\rho^- < \hat{\rho} < \rho^+$. The problem is solved the following way. Let $\nu \equiv \hat{\rho}(0)$ for our potential solution. Then the maximum principle again says that $\hat{\rho} > \nu$ for $y > 0$ and $\hat{\rho} < \nu$ for $y < 0$. Therefore if we define

$$G_\nu(\hat{\rho}) = \begin{cases} G^+(\hat{\rho}) & \text{for} \quad \hat{\rho} > \nu \\ G^-(\hat{\rho}) & \text{for} \quad \hat{\rho} < \nu, \end{cases}$$

(2.17) becomes

$$\frac{1}{2}\,\partial_y^2\hat{\rho} + G_\nu(\hat{\rho}) = 0.$$

From (2.18), $\partial_y\hat{\rho} = 0$ when $\rho = \rho^{\pm}$. Multiply the above by $\partial_y\hat{\rho}$ and integrate to obtain

$$\frac{1}{2}(\partial_y\hat{\rho})^2 + 2\int_{\rho^-}^{\hat{\rho}} G_\nu(u)du = 0. \tag{2.19}$$

Setting $\hat{\rho} = \rho^+$, $\partial_y\hat{\rho} = 0$, we get

$$\int_{\rho^-}^{\rho^+} G_\nu(u)du = 0. \tag{2.20}$$

This is a necessary condition for the existence of a solution of (2.17), (2.18). It turns out to be sufficient as well. Since $G_\nu(\rho)$ changes sign at $\rho = \nu$, ν may be chosen in a unique manner to satisfy (2.20). Further integrating (2.19) gives the whole solution $\hat{\rho}(y)$.

So a solution $\hat{\rho}(y)$ does exist; it varies from near ρ^- to near ρ^+ in a ξ-interval $\Delta\xi = O(s)$ (the transition layer). With this layer joining ρ^- to ρ^+, and $\rho(\xi)$ set equal to $\rho_c(\xi)$ outside the layer, we have constructed a suitable approximation to (2.13c). It is no longer discontinuous, but does make a sharp transition.

It now appears that in place of (2.16), we should solve the equation analogous to (2.16) but with ρ_c replaced by the transition layer approximation for ρ. There is also still the question as to whether the terms neglected in (2.13b) are really small. Regarding the first point, it turns out to be allowable to solve (2.16) with the original discontinuity ρ_c rather than the smoothed out version. This is because the difference between the two functions, though not uniformly small, is small except for a short (in the scale of ξ) transition layer, and a difference in a small interval does not affect the solution p, to the lowest order approximation. This fact is seen to be plausible by writing (2.16), which takes the form

$$\frac{1}{2}\partial_\xi(\rho_c^2\Delta^2\partial_\xi p) - \rho_c^2\Delta\mu\partial_\xi p = -\rho_c^2\Delta f, \tag{2.21}$$

as an integral equation, using $p(-\infty) = 0$, $p(\infty) = 1$ (or vice versa, depending on which way the cline slopes).

At this stage one point should be clarified. Does the discontinuity in ρ_c affect the very existence of a solution of (2.21)? The answer is no; (2.21) is a reasonable equation to have for a model. What the discontinuity does is tell us that the solution p will have a discontinuous derivative at $\xi = 0$, and the discontinuity will be such that $\rho_c^2\Delta^2\partial_\xi p$ is continuous there. This can be seen by integrating (2.21) with respect to ξ from $-\varepsilon$ to ε, then letting $\varepsilon \to 0$. This fact is standard in the theory of ordinary differential equations. Nagylaki (1976) obtained the same condition on $\partial_\xi p$, when studying discontinuous Δ. See also Sawyer (1978a).

With (2.21) established, we have to consider whether the discarded terms in (2.13b) are small. But since, in our approximation, $p_{12} = p(1 - p)$, the second derivative in (2.13b) becomes

$$\partial_\xi(\rho_c^2\Delta^2\partial_\xi p(1 - p)) = \partial_\xi(\rho_c^2(1 - 2p)\Delta^2\partial_\xi p) = (1 - 2p)\partial_\xi(\rho_c^2\Delta^2\partial_\xi p) - 2(\partial_\xi p)(\rho_c^2\Delta^2\partial_\xi p).$$

Since p satisfies (2.21), the quantities $\partial_\xi(\rho_c^2\Delta^2\partial_\xi p)$ and $\partial_\xi p$ are bounded (though perhaps discontinuous). Therefore the terms neglected in (2.13b) are really $O(s)$.

To summarize, we have found that in the case of a discontinuous carrying capacity ρ_c, the cline solution we got by solving (2.21) is still legitimate, in the following sense: Knowing p, if we now set $p_{12} = p(1 - p)$, and set ρ equal to the discontinuous ρ_c smoothed as indicated above in a transition layer, then the three functions p, p_{12}, and ρ will satisfy (2.13) except for terms of the order s.

Needless to say, the same type of analysis will show that discontinuities in V and M are also allowable within the framework of our model (2.16).

2.7 Discussion

Our aim has been to examine the degree of generality with which selection-migration phenomena can be modeled by a single nonlinear diffusion equation, and to provide a more satisfactory justification for such modeling in the cases which have been studied previously. Fisher (1937) was the first to use such a model, and we call generalizations of his original equation by his name. Our generalization (2.16) may account for nonuniformities in carrying capacity of the environment, in diffusivity, and in selection pressures, as well as for density dependent selection.

We have gone into the formal justification of (2.16), under our originally stated hypotheses, and found it to be correct if the heterogeneities in the given data have characteristic spatial or temporal variation compatible with the variables ξ and τ. We have also indicated that it is correct under other circumstances as well, and the example of a carrying capacity discontinuous in x was analyzed. Many other examples in which the data depend discontinuously on x or t could be similarly analyzed. If the birth or death rate (hence carrying capacity) has a temporal discontinuity, for instance, but is otherwise independent of t, and clines are being considered, then immediately after the discontinuous change in the rate takes place, there will be a readjustment period of length $\Delta\tau \sim s$ ($\Delta t \sim 1/\bar{r}$). After this period, a new cline will stabilize. Such a time interval is an "initial layer," and the readjustment process may be examined on the basis of (2.13) by using a shorter time scale, suited to the duration of the process.

When the heterogeneities vary on scales which are finer (in space and/or time) than those of ξ and τ, for example if ρ_c is a rapidly varying function relative to those variables, then it is probable that an equation of Fisher type may still legitimately be derived by a homogenization process, in the sense of Bensoussan, Lions, and Papanicolaou (1978).

3. FORMULATION OF MATHEMATICAL PROBLEMS

With a specific reaction-diffusion system at hand, one generally wishes either to find a solution satisfying certain subsidiary conditions, or to determine certain qualitative properties shared by many solutions. The various possibilities here generate a variety of mathematical problems, the most important of which are described in this chapter.

The broadest framework within which we shall operate is the following generalization of (1.20) to more than one space dimension (to conform more with convention, we shift notation from $\underline{\rho}$ to $\underline{u}$):

$$\frac{\partial u_i}{\partial t} = L_i u_i + F_i(x,t,\underline{u}), \quad i = 1, \dots, n, \tag{3.1}$$

where the L_i are uniformly elliptic differential operators in the variables $x = (x_1, \dots, x_n)$, of the form

$$L_i u \equiv -\Sigma \frac{\partial}{\partial x_k} H_{ik}, \quad -H_{ik} = \sum_{\ell} D_{ik\ell}(x,t) \frac{\partial u}{\partial x_\ell} + C_{ik}(x,t)u.$$

3.1 The standard problems

Let Ω be a domain in m-space (we often take $m = 1$, in which case Ω is an interval). When $m > 1$ and when Ω is not the whole space, we assume its boundary $\partial\Omega$ to have a unit normal which is a smooth function of the position on $\partial\Omega$.

For $T > 0$, let $Q_T = \Omega \times (0,T)$.

Initial boundary value problems: Find a continuous function $u: \bar{Q}_T \to R^n$, bounded in x for each fixed $t \in (0,T)$, the derivatives appearing in (3.1) existing and continuous in $\Omega \times (0,T]$, and satisfying

(i) (3.1) in Q_T,

(ii) initial condition $u(x,0) = \phi(x)$, where ϕ is a prescribed vector function, bounded and continuous on $\bar{\Omega}$,

(iii) boundary conditions on $\partial\Omega \times (0,T]$, of a type described in Section

1.3 for the case $m = 1$. Specifically, we have

$$u_i(x,t) = c_i(x,t) \qquad \text{(Dirichlet condition)} \tag{3.2a}$$

$$\sum_k \nu_k(x) H_{ik} = a_i(x,t) \qquad \text{(Neumann condition)} \tag{3.2b}$$

$$\sum_k \nu_k(x) H_{ik} = a_i(x,t) + b_i(x,t)u \qquad \text{(Robin condition).} \tag{3.2c}$$

Here $\underline{\nu}(x) = (\nu_1, \ldots, \nu_m)$ is the unit outward normal at x, and the $b_i > 0$.

Initial value problem: The same, with $\Omega = R^m$, and no boundary conditions.

Boundary value problems: Find a bounded function $u: \bar{\Omega} \times (-\infty,\infty) \to R^n$, continuous together with the derivatives appearing in (3.1), satisfying (3.1) in $\Omega \times (-\infty,\infty)$, and one of the three boundary conditions on $\partial\Omega \times (-\infty,\infty)$.

Stationary problems: Find a time-independent solution of the boundary-value problem, together with one of the boundary conditions (3.2); the prescribed functions and coefficients are now supposed to be time-independent.

Periodic problems: Supposing the functions in (3.1) and (3.2) are periodic in time with common period, find a solution of a boundary value problem which is periodic in time with the same period.

In most of what we shall do, the question of stability will be an important, though usually not an easy, question. Most of the existing results pertain to C^o-stability. Accordingly, we define

$$|u(\cdot,t)|_o \equiv \sup_{x \in \Omega; i} |u_i(x,t)|.$$

Definition: Let $u(x,t)$ be a solution of (3.1) on $\Omega \times (0,\infty)$, satisfying a boundary condition of a type (3.2). It is C^o-stable if, given $\varepsilon > 0$, there is a $\delta > 0$ such that every solution ρ^* of (3.1) on $\Omega \times (0,T)$ for some $T > 0$ satisfying the same boundary condition and satisfying $|u^*(\cdot,0) - u(\cdot,0)|_o < \delta$, (i) can be continued to be a solution of (3.2), with the same boundary condition, on $\Omega \times (0,\infty)$, and (ii) $|u^*(\cdot,t) - u(\cdot,t)|_o < \varepsilon$ for all $t > 0$.

Definition: Let $u(x,t)$ be a C^o-stable solution as in the definition above. It is C^o-asymptotically stable if there is a $\delta > 0$ such that every solution u^* with the described properties also satisfies

$$\lim_{t\to\infty} |u^*(\cdot,t) - u(\cdot,t)|_o = 0.$$

Asymptotic stability in the presence of symmetries is an important concept. It will be defined below.

3.2 Asymptotic states

If only the long time behavior of solutions is of importance to the investigator, then the concept of an asymptotic state is significant. The asymptotic state of a solution specifies its ultimate behavior while ignoring transient effects. It could therefore be defined as a collection of solutions, all approaching the same solution as $t \to \infty$. When there are symmetries in the reaction-diffusion system under consideration, however, a whole class of solutions can immediately be generated from any given solution, by subjecting it to an appropriate group of transformations. All these other solutions, in a sense, have the same asymptotic behavior, so should belong to the same asymptotic state.

To be specific, we consider now the case $L_i u \equiv D_i \Delta u$, where D_i are non-negative constants and Δ is the Laplacian, and F is independent of (x,t),

$$\frac{\partial u_i}{\partial t} = D_i \Delta u_i + F_i(u). \tag{3.3}$$

In terms of modeling, this corresponds to a homogeneous isotropic medium with no drift.

In comparing solutions, it really should not matter if one is displaced from another in space or time. More generally, let G be the group of transformations on (x,t) generated by rigid motions and reflections in x, and translations in t. Clearly these transformations leave the set of solutions of (3.3) invariant. If $T \in G$, we note by Tu the solution obtained from u by operating on (x,t)

by the transformation T. We shall consider u and Tu to be equivalent. For particular systems (3.3) with added symmetry properties, the group G can be extended, perhaps, to act on the vector u as well. In such cases, it may be desirable to use the larger group in defining equivalence. On the other hand, for systems whose coefficients depend on x (for example), or which contain first order derivatives, G will have to be reduced to a smaller group.

Definition: Let $u^{(1)}$, $u^{(2)}$ be two solutions of (3.3) for all x and for large enough t. They are asymptotically equivalent if, for some $T \in G$,

$$\lim_{t\to\infty} \sup_x |u^{(1)}(x,t) - Tu^{(2)}(x,t)| = 0.$$

Definition: An asymptotic state for (3.3) is an (asymptotic) equivalence class of solutions.

The class asymptotically equivalent to a given u is denoted by [u]. If ϕ is a bounded continuous function of x, we denote by u_ϕ the solution of the initial value problem with initial data ϕ.

We take the stance that the stable asymptotic states (SAS's) are the important ones, as they are the ones generally seen in applied contexts. Again, we define stability in the context of the space C^o, because that is where most existing stability results for RD systems reside.

Definition: An asymptotic state [u] is C^o-stable if the set

$$\{\phi : u_\phi \in [u]\}$$

is open in $C^o(\mathbb{R}^n)$.

Implicit in this definition is that for ϕ in this set, u_ϕ exists globally. This set is nonempty; in fact it contains the functions $u(\cdot,t)$ for all large enough t. SAS's may be represented by solutions which can be perturbed by a uniformly small function without destroying their long-time behavior.

In the case of stationary or plane wave solutions, another definition for stability is in terms of the L_∞ spectrum of the linearization of the right side of (3.3) about the solution in question (the problem should be cast in a moving coordinate frame in the case of plane waves). The criterion is that except for a simple eigenvalue at the origin (which always exists), the spectrum is in the left half plane and bounded away from the imaginary axis. Sattinger (1976a,b; 1977a) proved that in typical cases, stability in this sense implies C^0-stability.

The stable asymptotic states are usually the important ones, because small changes in the solution at any fixed time will not affect the solution type in the long run. If a certain solution belongs to an unstable asymptotic state, that means its appearance is only accidental, and that a small disturbance at some time will destroy its asymptotic nature. So it would probably rarely be observed in nature.

Unstable states are important in some contexts, however; for example, in Fisher's model of natural selection and waves of advantageous genes (1937), the small perturbation is an advantageous mutation, which would seldom occur, but has lasting effects and alters the asymptotic nature of the solution. This model was discussed in Chapter 2; the possibilities for stable and unstable asymptotic states will be considered further in Chapter 4.

The following question is central in the theory of reaction-diffusion equations:

Question: Given a particular reaction-diffusion system, what are its stable asymptotic states?

Definition: A solution $u(x,t)$ is of "permanent type" if it is defined, and is a solution, for all real t, negative as well as positive.

The equation (3.3) defines a semi-flow on a suitable set $S \subset C^0$, in that given any $\phi \in S$ and $t \geq 0$, there is a function $\psi \in S$ with $\psi(x) = u_\phi(x,t)$, and these solution operators form a semigroup with parameter t. However, on the range $\hat{S}$ of a solution of permanent type, (3.3) defines a flow, because now t can be either negative or positive.

The following are familiar examples of solutions of permanent type:

- solutions periodic in t
  - x-independent oscillating solutions
  - wave trains (see below)
  - target patterns $(u(x,t) = U(|x|,t)$, U periodic in $t)$
  - rotating spiral patterns $(n = 2$, $x = r(\cos\theta, \sin\theta)$, $u(x,t) = U(r,\theta - ct)$, U periodic in second argument)
- traveling waves (solutions of the form $u(x,t) = u(x - ct)$, c a velocity vector)
  - stationary solutions $(c = 0$, u time-independent)
  - plane waves $(u(x,t) = U((x - ct)\cdot\nu) = U(x\cdot\nu - |c|t)$, ν a unit vector in the direction of $c)$
    - wave trains (U periodic)
    - wave fronts (U monotone and bounded)
    - pulses $(U(-\infty) = U(\infty)$, U not constant).

Question: What are the stable asymptotic states of permanent type? Or more specifically, of one of the above types?

This should be an easier question than the preceding, because here one seeks only solutions with certain properties.

Other solutions, not of permanent type, such as diverging pairs of fronts and expanding circular or spherical fronts, have been studied.

Recognizing that traveling waves cannot exist forever in a bounded domain, but may represent solution forms which appear stable and permanent for relatively long time durations, Barenblatt and Zel'dovich (1971) categorize them as "intermediate asymptotic" solutions.

Associated with (3.3) is the corresponding system of ordinary differential equations obtained by eliminating all spatial derivatives:

$$\frac{du}{dt} = F(u). \tag{3.4}$$

It is called the kinetic system corresponding to (3.3). The concept of asymptotic state is valid for it as well. Solutions of (3.4) can be considered as space-independent solutions of (3.3) as well, and therefore form still another category of solutions of (3.3).

Question: What relation is there between the stable asymptotic states (SAS's) of (3.3), and those of (3.4)?

As an example, let us consider the scalar equation in one space variable,

$$u_t = u_{xx} + f(u). \tag{3.5}$$

If f has only a single zero at $u = u_o$, and changes sign there from positive to negative (as in the figure below), then it is easily seen that the corresponding

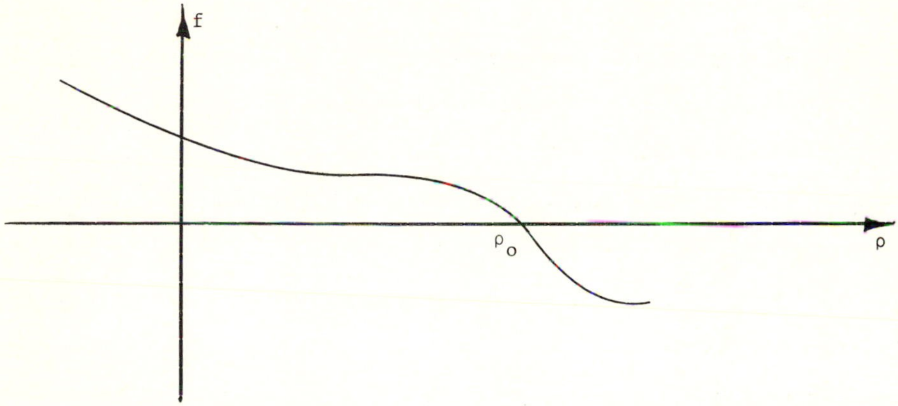

kinetic equation $du/dt = f(u)$ has u_o as stationary value, and it is globally stable. In fact, if $u(0) < u_o$, u will increase with time and approach u_o as $t \to \infty$. The opposite happens if $u(0) > u_o$. Thus, (3.4) has a single SAS. It can (and will later) be shown that $[u_o]$ is also the only SAS of (3.5). In fact, it is generally true that for scalar equations, every stable rest point of (3.4) is a stable solution of (3.5).

Now suppose f has two stable zeros, hence an unstable one between them, as shown below. For simplicity, let the zeros be 0, α, and 1. (When f has two or more zeros, (3.5) is sometimes called Fisher's equation; see Chapter 2.) So now (3.5) has two constant stable asymptotic states, $[0]$ and $[1]$. It turns out to have nonconstant ones as well, in the form of traveling wave fronts which approach

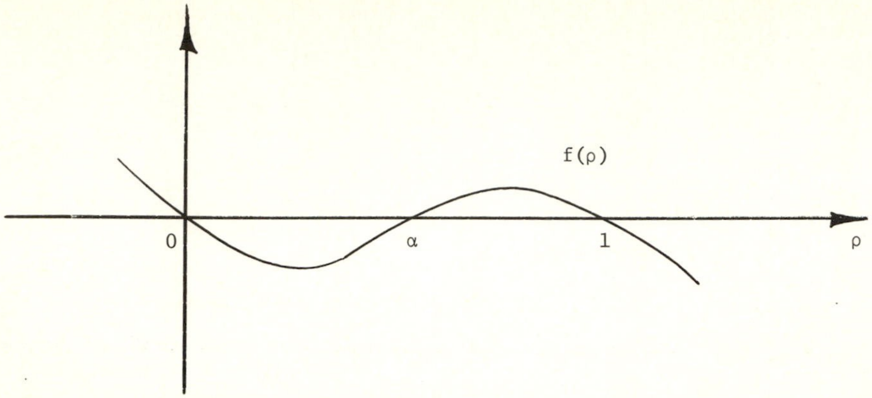

one of the stable states as $x \to -\infty$ and the other as $x \to \infty$. Specifically, there is a constant c and a function $U(z)$, with $U(-\infty) = 0$, $U(\infty) = 1$, such that $u(x,t) = U(x - ct)$ and $u(x,t) = U(-x - ct)$ are stable solutions of (3.5) (we shall prove this in Section 4.5). One of these solutions is the reflection of the other, so they both represent the same SAS.

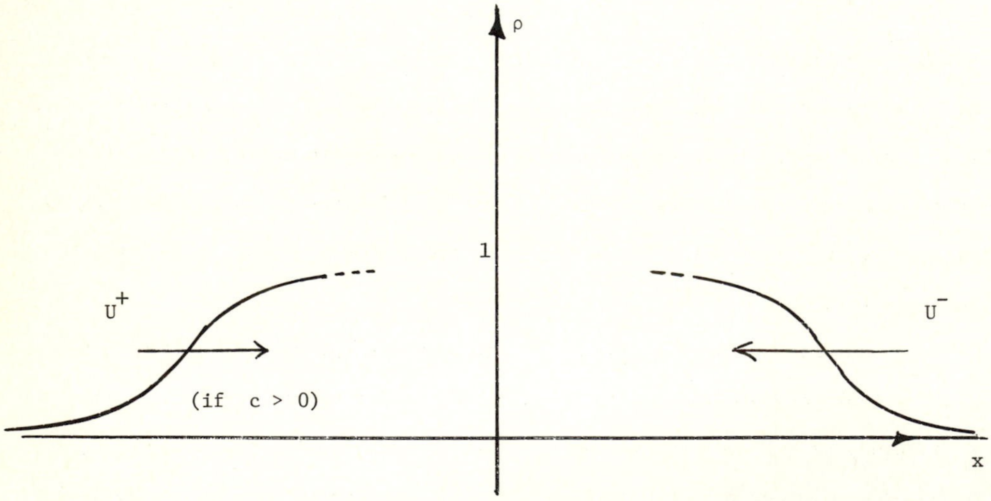

This makes three stable asymptotic states of permanent type. The question is, are there others? It can be shown that there are no others in the form of traveling waves. I conjecture that there are no others of permanent type. But if $c \neq 0$, there is another, not of permanent type. It takes the form of two diverging wave fronts, as shown:

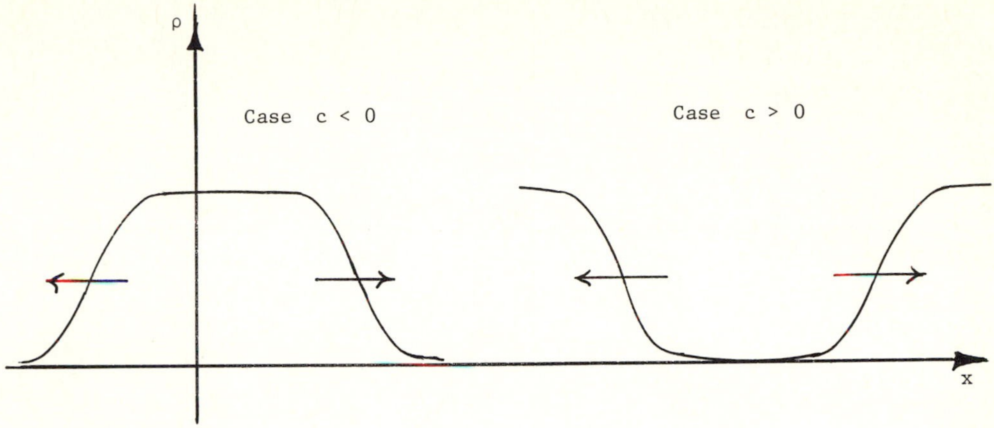

Call this asymptotic state [W]. There are still other stable asymptotic states formed from an infinite set of such diverging pairs of fronts, the pairs spaced ever further apart as $|x| \to \infty$. But the following is probably true.

Conjecture: If $\phi(x)$ is bounded away from α for large $|x|$, then $u_\phi \in [0], [1], [U], [W]$, or u_ϕ is in an unstable asymptotic state.

Under the stated assumption, a conclusion which is close to this has been established (Fife 1979, building on Fife and McLeod 1976 and Aronson and Weinberger 1974). Specifically, it was shown that u_ϕ will belong to one of five possible categories, four of them being the SAS's listed. Solutions in the fifth category were characterized explicitly and shown to be not uniformly stable; they are probably unstable.

The appearance of wave fronts, as above, suggests one possible way of constructing stable asymptotic states for the reaction-diffusion system (3.3) (at least in the case of one space variable) from two known SAS's of (3.4): Find a solution which, for each t, behaves as $x \to -\infty$ like one of the given stable states of (3.4), and as $x \to \infty$ like the other. When $n = 1$ (3.5), the two given states must be constants; then the construction would give a wave front joining them. (But it is not always possible to join them (Fife and McLeod 1976) as in the example above.) Some examples are known for wave fronts like this, for $n > 1$ (Conley and Fife 1979, Murray 1976, Fife 1976c). But other possibilities also suggest themselves, like joining a periodic solution to a constant solution. Some results along similar lines were obtained by Feinn and Ortoleva (1977); but more work needs to be done.

Another approach to constructing new SAS's of (3.3) from a known SAS of (3.4) applies when the latter is periodic in time. Then by a perturbation procedure, it is usually possible to obtain wave train solutions of (3.3) with arbitrarily long wave length, which locally oscillate in the manner of the known solution (Howard and Kopell 1973; Section 8.1).

3.3 Existence questions

Asymptotic states were defined in Section 3.2 in terms of solutions existing for all $t > 0$. It is worthwhile knowing whether there does exist such a global solution, for given initial data ϕ. An easy counterexample can be provided:

$$\frac{du}{dt} = u^2, \quad u(0) = 1.$$

The solution is $u(t) = 1/(1 - t)$, which exists only for $t < 1$, and becomes unbounded as $t \to 1$. The fact is, the only way a solution of (3.3) (or, a fortiori of (3.4)) can cease to exist at some finite time, is for it to become unbounded. Therefore if we can obtain some a priori bound for the solution, we can infer global existence. Such a priori bounds can, in fact, be obtained in important situations. In the case of the ordinary differential equations (3.4), see, for example, Theorem 1.4. Other cases will be treated, for reaction-diffusion systems and scalar equations, in later sections. The basic existence theorem is the following. Its proof will not be given, but is fairly standard.

THEOREM 3.1. Assume F is a continuously differentiable function of u. Let $\phi \in C^0(R^n)$ be given. Suppose any potential solution of the initial value problem (3.3); $u(x,0) = \phi(x)$ must satisfy

$$|u(\cdot,t)|_o \le K(t), \quad t \in R^+,$$

where K is some positive function. Then there exists such a solution u, and it is unique.

For the equations of ecology, natural situations occur for which such an a priori bound may be established. See, for example, the "food-pyramid" argument in Williams and Chow (1978).

Similar theorems hold for the various initial boundary value problems associated with (3.3). Probably the most complete global existence results are those of Amann (1977).

4. THE SCALAR CASE

Now we begin a study of the <u>scalar</u> equation

$$u_t = u_{xx} + f(u). \tag{4.1}$$

The following topics will be taken up:

Comparison and Lyapunov methods

Stationary solutions and their stability analysis

Wave front solutions and their stability analysis

The inhomogeneous Fisher's equation and clines (briefly)

Aronson (1976) and Diekmann and Temme (1976) provided surveys with slightly different emphases. We shall always assume that f is continuously differentiable for all values of u.

4.1 Comparison methods

Methods based on maximum principles have been developed for scalar semilinear parabolic equations much more general than (4.1), with or without boundary conditions, and find wide application to a variety of problems (Protter and Weinberger 1967). So here, treating the initial value problem for (4.1), we are concerned with a rather special case.

<u>Strong maximum principle</u>: (See, for example, Protter and Weinberger (1967), Friedman (1964).) Let c be a bounded function in $\bar{Q}_T$, where $Q_T = (-\infty,\infty) \times (0,T)$. Let $v(x,t)$ be continuous in $\bar{Q}_T$, have derivatives v_t and v_{xx} continuous in Q_T, and satisfy

$$v_t - v_{xx} - c(x,t)v \geq 0 \quad \text{in} \quad Q_T,$$

$$v(x,0) \geq 0.$$

Then either

(i) $v > 0$ in Q_T, or

(ii) $v \equiv 0$ in Q_T.

Definition: $\underline{u}(x,t)$ is a regular subsolution of (4.1) in Q_T if it has the above continuity and differentiability properties, and

$$N[\underline{u}] \equiv \underline{u}_t - \underline{u}_{xx} - f(\underline{u}) \le 0.$$

Definition: $\underline{u}(x,t)$ is a subsolution of (4.1) in Q_T if $\underline{u}(x,t) = \max_{1 \le j \le q} u_j(x,t)$ for some collection of q regular subsolutions $\{u_j\}$. Supersolutions are defined in the same way.

THEOREM 4.1. Let $\underline{u}$ and $\bar{u}$ be sub- and supersolutions satisfying $\underline{u}(x,0) \le \bar{u}(x,0)$. Then

(i) $\underline{u} < \bar{u}$ in Q_T, or

(ii) $\underline{u} \equiv \bar{u}$.

(Note that either function could be a solution, and if alternative (ii) holds, they both must be.)

Proof: Let $\underline{u}_j$, $\bar{u}_k$ be a regular subsolution and supersolution with $\underline{u}_j(x,0) \le \underline{u}(x,0) \le \bar{u}(x,0) \le \bar{u}_k(x,0)$. Let $v = \bar{u}_k - \underline{u}_j$. Then

$$v_t - v_{xx} - c(x,t)v \ge 0,$$

where $c(x,t) = [f(\bar{u}_k(x,t)) - f(\underline{u}_j(x,t))]/[\bar{u}_k(x,t) - \underline{u}_j(x,t)]$ is bounded, since f is continuously differentiable. It follows from the Strong Maximum Principle that $\underline{u}_j < \bar{u}_k$ in Q_T, or else $\underline{u}_j \equiv \bar{u}_k$. In the latter case, we have that $\underline{u} \equiv \bar{u}$, which is alternative (ii) in the theorem. If that alternative is not true, we therefore have that $\underline{u}_j < \bar{u}_k$ for every pair $(\underline{u}_j, \bar{u}_k)$ with the properties indicated. Therefore

$$\underline{u}(x,t) = \max_j \underline{u}_j(x,t) < \min_k \bar{u}_k(x,t) = \bar{u}(x,t),$$

and alternative (i) holds.

The following result was used in Aronson and Weinberger (1975), and possibly prior to that paper as well.

THEOREM 4.2. Let $\underline{\phi}$ be a time-independent subsolution, and u an exact solution satisfying $u(x,0) = \underline{\phi}(x)$. Then either u is strictly increasing in t, for each fixed x, or $u \equiv \underline{\phi}$.

Proof: Assume $u \not\equiv \underline{\phi}$. By Theorem 4.1, we know that $u(x,t) > \underline{\phi}(x)$ for $t > 0$. For some small $\delta > 0$, let $u^{\delta}(x,t) = u(x,t+\delta)$. It is a solution of (4.1), hence a supersolution. We have that $u^{\delta}(x,0) > u(x,0)$. By Theorem 4.1 with $\underline{u} = u$, $\bar{u} = u^{\delta}$, we conclude that $u(x,t+\delta) = u^{\delta}(x,t) > u(x,t)$. Since δ was arbitrary, the conclusion follows.

THEOREM 4.3. Suppose there exist sub- and supersolutions $\underline{u}$ and $\bar{u}$ defined for all $t > 0$, bounded in x for each t, and satisfying $\underline{u}(x,0) \le \bar{u}(x,0)$. Let ϕ be a continuous function satisfying $\underline{u}(x,0) \le \phi(x) \le \bar{u}(x,0)$. Then there exists a unique solution of (4.1) satisfying $u(x,0) = \phi(x)$, and it satisfies

$$\underline{u}(x,t) \le u(x,t) \le \bar{u}(x,t).$$

Proof: Theorem 4.2 provides an a priori bound which can be used with Theorem 3.1 to establish the result.

If we consider solutions which are not defined for all x, but rather for x in some bounded interval $\bar{\Omega}$ (where we take Ω to be open), then analogs of Theorems 4.1-4.3 are true. The domain Q_T should then be redefined as $Q_T = \Omega \times (0,T)$. In the case of the analog of Theorem 4.1, an additional hypothesis relating $\underline{u}$ and $\bar{u}$ on the boundary $\partial\Omega \times (0,T)$ needs to be imposed. It can take one of the two forms

(a) $\underline{u} \le \bar{u}$,

(b) $\partial_{\nu}\underline{u} - a\underline{u} \le \partial_{\nu}\bar{u} - a\bar{u}$, where $a \ge 0$ and ∂_{ν} represents the derivative in the direction $(\pm x)$ outward from Ω. Then the conclusion still holds.

In the case of Theorem 4.2, we must require that u and $\underline{\phi}$ satisfy boundary inequalities of one of the types given above, with $\underline{u} = \underline{\phi}$, $\bar{u} = u$. In the case of Theorem 4.3, we likewise require that $\underline{u}$ and $\bar{u}$ satisfy one of the given boundary inequalities (a) or (b). Suppose that it is (a). Then if $g(x,t)$ is some smooth function satisfying $\underline{u}(x,t) \leq g(x,t) \leq \bar{u}(x,t)$ on $\partial\Omega \times (-\infty,\infty)$, there will exist a unique global solution of the initial-boundary value problem with initial data ϕ and Dirichlet data g. A similar result holds in the case of inequality (b).

4.2 Derivative estimates

Among the types of a priori estimates in partial differential equations, the Schauder estimates are among the most useful, particularly in nonlinear problems. They were developed by Schauder for second order elliptic equations in the 1930's, and later extended to other categories of equations. In particular, Schauder estimates for second order parabolic equations were derived by A. Friedman in the 1950's; see their full exposition in Friedman (1964). The results we give here are simple cases and consequences of the Schauder estimates.

Let Q be a rectangle $(x_o,x_1) \times (t_o,t_1)$ in the (x,t) plane, each of the numbers x_o, x_1, t_1 being either finite or infinite. Corresponding to Q and to a number $\delta > 0$, let Q_δ be the smaller rectangle $[x_o + \delta, x_1 - \delta] \times [t_o + \delta, t_1]$ (assuming $x_1 - x_o < 2\delta$, etc.). For a function u with derivatives appearing in (4.1) defined in Q, let

$$|u|_o^Q \equiv \sup_{(x,t) \in Q} |u(x,t)|, \qquad |u|_1^Q \equiv |u|_o^Q + |u_x|_o^Q,$$

$$|u|_2^Q \equiv |u|_1^Q + |u_{xx}|_o^Q + |u_t|_o^Q.$$

(We allow the possibility that these norms be equal to infinity.)

THEOREM 4.4. Let u and h satisfy $u_t - u_{xx} = h(x,t)$ in Q. Then for some $C > 0$, depending only on δ and not on u, Q, or h,

$$|u|_1^{Q_\delta} \leq C(|h|_o^Q + |u|_o^Q). \tag{4.2a}$$

Furthermore if h_x exists in Q, then

$$|u|_2^{Q_\delta} \leq C(|h|_1^Q + |u|_o^Q), \tag{4.2b}$$

and the moduli of continuity of u_{xx} and u_t in Q_δ are bounded by a modulus depending only on δ. Let Q'_δ be the rectangle $[x_o,x_1] \times [t_o + \delta,t_1]$, and $B_{\delta/2} = \{x_o,x_1\} \times (t_o + \delta/2,t_1]$ (the lateral boundary of $Q'_{\delta/2}$). If, in addition to the above, u, restricted to $B_{\delta/2}$, is twice differentiable in t, then

$$|u|_2^{Q'_\delta} \leq C(|h|_1^Q + |u|_o^Q + |u_{tt}|_o^{B_{\delta/2}}). \tag{4.3}$$

The estimate (4.2a) follows from Friedman's $(1 + \delta)$-estimate (see his text (1964)); (4.2b) and (4.3) follow from the Schauder estimates of interior type and boundary type, respectively.

THEOREM 4.5. Let u satisfy (4.1) in Q. Then for some C depending only on δ,

$$|u|_1^{Q_{\delta/2}} \leq C(|f \circ u|_o^Q + |u|_o^Q), \tag{4.4a}$$

$$|u|_2^{Q_\delta} \leq C(|f' \circ u|_o^Q |u|_1^{Q_{\delta/2}} + |u|_o^Q). \tag{4.4b}$$

If u is continuous in $Q \cup B_{\delta/2}$ and u, restricted to $B_{\delta/2}$, is twice differentiable in t, then for some C depending on δ, $|f \circ u|_o^Q$, and $|f' \circ u|_o^Q$,

$$|u|_2^{Q'_\delta} \leq C(1 + |u|_o^Q + |u_{tt}|_o^{B_{\delta/2}}). \tag{4.5}$$

In either case, the moduli of continuity of u, u_x, u_{xx}, and u_t in Q_δ or Q'_δ, respectively, are bounded by a modulus depending only on f and δ.

This follows from Theorem 4.4 by setting $h = f \circ u$.

Derivative estimates such as these, which incidentally are also true for systems of reaction-diffusion equations, are important in establishing the "compactness of orbits" for solutions of such equations and systems. Compactness is needed, for example, in Lyapunov methods, as we shall see. The main compactness statement takes the form of the following theorem.

THEOREM 4.6. Let $u(x,t)$ satisfy (4.1) for all $t > 0$, and $|u(x,t)| \leq M$ for some constant M. Then given any sequence $\{t_n\}$ with $t_n \to \infty$ as $n \to \infty$, there exists a subsequence $\{t_n'\}$ and a function $w(x)$ such that $u(x,t_n') \to w(x)$, $u_x(x,t_n') \to w_x(x)$, $u_{xx} \to w_{xx}$ as $n \to \infty$, these convergence processes being uniform for x in bounded intervals.

Proof: In Theorem 4.5, take Q to be the whole $\frac{1}{2}$-space $\{t > 0\}$. Then $|u|_0^Q \leq M < \infty$, so by (4.4) and the final statement of the theorem, the functions $u(x,t_n)$, $u_x(x,t_n)$, and $u_{xx}(x,t_n)$ all have a common bound, and their moduli of continuity as well have a common bound, independent of n. By Arzela's theorem, there is a subsequence of each of them which converges uniformly on bounded intervals. Moreover, the functions to which u_x and u_{xx} converge must be the corresponding derivatives of the function w to which u converges.

Corresponding convergence statements, again based on Arzela's theorem and Theorem 4.5, hold for solutions of boundary value problems on finite x-intervals, provided the functions arising in the boundary conditions are sufficiently regular.

4.3 Stability and instability of stationary solutions

In this section we shall construct all stationary solutions of (4.1), and analyze the stability of the bounded ones by means of the comparison methods developed in Section 4.1. We shall find that for problems on the entire real line, the only stable ones are the constants and the monotone ones. In the case of the stable constant solutions, estimates can be made on their domains of attraction. In the following section, it will be shown that wave fronts are also stable when they connect two stable rest states, and again, estimates can be made for their domains of attraction.

The material in this section was taken from Fife (1977a). Hagan (1979) has used comparison methods to establish instability results for many nonmonotone solutions of many types of scalar equations and generalizations of RD systems, incorporating dependence on past history and various types of nonlinearities. At the same time, he showed that monotone traveling waves are generally stable, in some sense or another.

We begin with a construction of the time-independent solutions of (4.1) defined for all x, i.e., solutions $\psi(x)$ of

$$\psi'' + f(\psi) = 0, \quad x \in (-\infty,\infty). \tag{4.6}$$

First, it is clear that any solution remains a solution when subjected to a shift in the independent variable, $\psi(x) \to \psi(x - x_o)$, or to a sign change in the independent variable: $\psi(x) \to \psi(-x)$. We shall view the new solutions so produced as not being essentially different from the original one.

Secondly, (4.6) may be integrated once to obtain the equivalent equation

$$\frac{1}{2}(\psi')^2 + V(\psi) = E = \text{const}, \tag{4.7}$$

where the "potential" $V(\psi) \equiv \int_0^\psi f(s)ds$. Indeed, if we multiply (4.6) by ψ', we notice that the left side becomes the derivative of the left side of (4.7). (If x were replaced by a time variable, (4.6) would represent the equation of linear motion of an object in a force field f; then (4.7) is a statement that the energy is constant.)

Clearly every constant solution of (4.6) is a zero of $f = V'$. The nonconstant ones can be characterized in the following way.

We consider the graph of the function $V = V(u)$ in the $u - V$ plane, and also the set S of horizontal line segments of nonzero length in that plane with the properties

(a) the segment lies strictly above the graph, except at its endpoints;

(b) each finite endpoint must lie on the graph.

Examples of such segments in S are in the figure below.

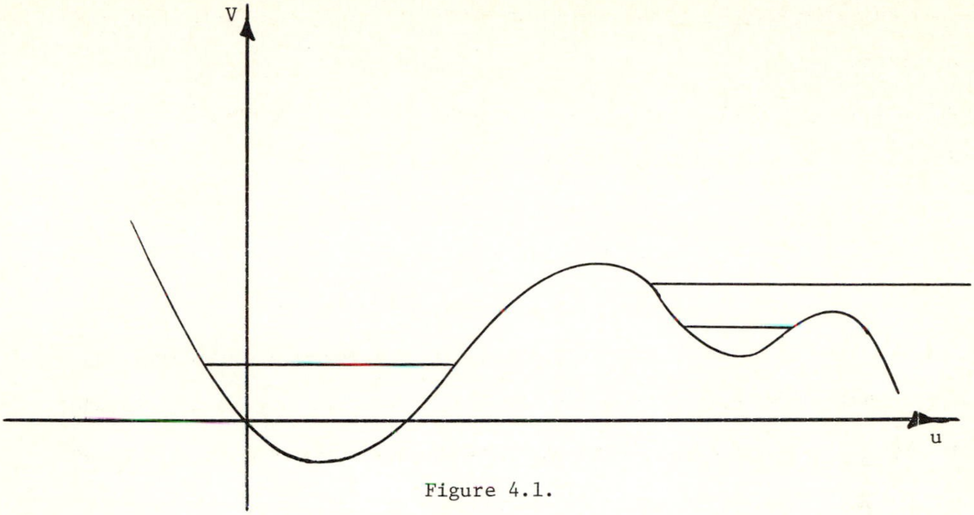

Figure 4.1.

THEOREM 4.7. There is a one-one correspondence between segments in S and the nonconstant solutions of (4.6), modulo shifts in x and reversals of sign. The correspondence is such that the segment in S overlays an interval on the u-axis which is the range of the corresponding solution.

Proof: Given a segment in S, let E be its ordinate, let I be the interval in u overlain by the segment, and let u_o be any point in the interior of I. Integrate the equation

$$\psi' = \sqrt{2(E - V(\psi))}, \qquad \psi(0) = u_o \tag{4.8}$$

to obtain a function $\psi(x)$ which is monotone increasing (since the right side of (4.8) is positive), and satisfies (4.6) (via (4.7)). There is a maximal interval (x_o, x_1) (it could be infinite) with $x_o < 0 < x_1$ on which the solution ψ exists. If $x_1 < \infty$, then either ψ becomes unbounded as $x \uparrow x_1$, or $V(\psi) \to E$ as $x \uparrow x_1$. In the latter case, from (4.8), $\psi'(x) \to 0$ as $x \uparrow x_1$. If $x_1 = \infty$ and ψ is bounded, then by monotonicity, ψ approaches some limit u_m, and $\psi' \to 0$, as $x \to \infty$. From (4.7), we see that $V(\psi) \to E$ in this case as well. Two cases could therefore occur, with reference to the right hand endpoint:

(1) $\lim_{x \uparrow x_1} \psi(x) = \infty$,

(2) $\lim_{x \uparrow x_1} \psi'(x) = 0$, $\lim_{x \uparrow x_1} V(\psi) = E$.

The first case happens if and only if $V(\psi) < E$ for all $\psi > u_o$; but this is precisely when the segment in question extends to infinity to the right. In the second case, the range of ψ extends upwards to the first value u_m at which $V(u_m) = E$. This is also the extent of the segment in question. The situation is similar for negative x, as $x \downarrow x_o$. Therefore in all cases, the range of the solution ψ obtained is I. In case (2) when $x_1 < \infty$, the function ψ may be extended beyond x, as an even function with respect to x_1: $\psi(x) = \psi(2x_1 - x)$. When we do so, the extended function is still a solution of (4.6), with a relative maximum at x_1. If $x_o > -\infty$ and ψ is bounded there, it may be extended to the left of x_o as well; and continued extensions this way will yield a periodic solution. Thus, to each segment in S, there corresponds a solution of (4.6) with the same range.

Now let ψ be a given nonconstant solution, and let u_o be in the interior of its range I (which must be an interval, of course). To ψ is associated a constant E in (4.7). It follows from (4.7) that $V(\psi) \leq E$ for $\psi \in I$, and that $V(\psi) = E$ for ψ equal to any maximum or minimum, hence for ψ at any finite endpoint of I. It follows that I is precisely the interval overlain by a horizontal line segment with ordinate E, covering u_o, and with finite endpoints (if there are any) on the curve $V(u) = E$. To complete the proof we merely have to show that $V(\psi) < E$ except at the endpoints. If this were not true, there would be a number u_1 in the interior of I, at which $V(u_1) = E$ and $V'(u_1) = f(u_1) = 0$. Then for ψ near u_1, $E - V(\psi) \sim K(u_1 - \psi)^2$, and as $\psi \uparrow u_1$, we may integrate (4.7) to obtain (for some $u_2 < u_1$ and some x_o)

$$x - x_o = \pm \int_{u_2}^{\psi} \frac{ds}{\sqrt{2(E - V(s))}} \sim \pm C \int_{u_2}^{\psi} \frac{ds}{u_1 - s} \to \pm\infty.$$

Thus the value u_1 may only be attained by ψ at $\pm\infty$, so the range of ψ cannot contain u_1 in its interior. This completes the proof.

The qualitative properties of solutions corresponding to representative examples of segments are listed below.

(a) Single maximum; $\psi \to \infty$ as $|x| \to \infty$.

(b) Single maximum; $\psi \to \psi_0 < \infty$ as $|x| \to \infty$.

(c) Periodic solution.

(d) Monotone solution with distinct finite limits as $x \to \pm\infty$.

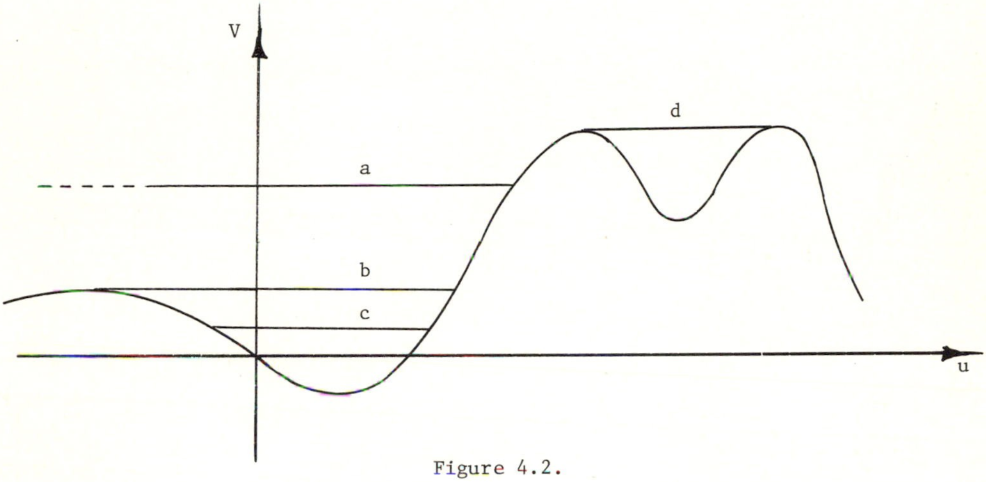

Figure 4.2.

As a general rule, an endpoint lying on a maximum point of the graph corresponds to an extreme value of u attained at $x = \pm\infty$, whereas one lying on the graph at a point where $V' \neq 0$ corresponds to an extremum of ψ attained at a finite value of x.

We now consider the stability of the various solutions. Our principal tool is the following consequence of the comparison theorems.

THEOREM 4.8. Let $\underline{\phi}$ be a uniformly continuous bounded subsolution of (4.1).

(a) If there exists a supersolution $\bar{\phi} \geq \underline{\phi}$, then there exists an exact solution $w(x)$ with the property that $\underline{\phi}(x) \leq w(x) \leq \bar{\phi}(x)$, and any time-independent exact solution $w_1(x) \geq \underline{\phi}(x)$ satisfies $w_1(x) \geq w(x)$ for all x. If $w(x) \geq \underline{\phi}(x) + \delta > \underline{\phi}(x)$, then $w(x)$ is a constant. If $\underline{\phi} \leq \phi \leq w$ and $u(x,t)$ is the solution of (4.1) with initial data $u(x,0) = \phi(x)$, then $\lim_{t\to\infty} u(x,t) = w(x)$, uniformly on bounded intervals in x.

(b) If there exists no supersolution $\bar{\phi} \geq \underline{\phi}$, and $u(x,t)$ is any solution of (4.1) with $u(x,0) \geq \underline{\phi}(x)$, then $\lim_{t \uparrow T} u(x,t) = \infty$ for some $T \leq \infty$.

Proof: Consider first case (a). Let v satisfy (4.1), with $v(x,0) = \underline{\phi}(x)$. By Theorem 4.2, v is strictly increasing, unless $v \equiv \underline{\phi}$ is an exact solution, in which case the conclusions of the theorem hold trivially. Also by Theorem 4.1, $v \leq \bar{\phi}$ for all t, so is bounded for fixed x. Therefore v approaches a limit $w(x)$ as $t \to \infty$. By Theorem 4.6, $v_x \to w_x$ and $v_{xx} \to w_{xx}$. By the monotonicity of v, we also know that $v_t \to 0$. Therefore passing to the limit in (4.1) (u replaced by v), we obtain $w_{xx} + f(w) = 0$. Now if w_1 is any other time-independent solution $\geq \underline{\phi}$, Min $[w_1,w]$ will be a supersolution larger than v, so that $\lim_{t\to\infty} v = w \leq \text{Min}\,[w_1,w] \leq w_1$, and hence $w_1 \geq w$.

Now suppose $w \geq \underline{\phi} + \delta$. Then for small ε, $w_\varepsilon(x) \equiv w(x + \varepsilon) \geq \underline{\phi}$ is a solution, so by the above, $w_\varepsilon \geq w$: $w(x + \varepsilon) \geq w(x)$. Replacing ε by $-\varepsilon$ and x by $x + \varepsilon$, we have $w(x) \geq w(x + \varepsilon)$, so $w(x) \equiv w(x + \varepsilon)$. Since ε was arbitrary, we have $w \equiv$ const. This fact was noted by Warren Ferguson.

Again by Theorem 4.1, the solution u in the statement of the present theorem satisfies $v \leq u \leq w$, so $\lim_{t\to\infty} u = w$.

Now consider case (b). Let $\phi_m = \sup \underline{\phi}(x)$. Clearly $f(u) > 0$ for $u > \phi_m$, since otherwise $\bar{\phi}(x) \equiv u_1 \geq \phi_m$ would be a supersolution. Let $K > \phi_m$, and define a new function $f_K \leq f$ so that $f_K(u) \equiv f(u)$ for $u \leq K$, $f_K(K + 1) = 0$. Let $u_K(x,t)$ satisfy (4.1) with f replaced by f_K, and $u_K(x,0) = \phi(x)$. Then $u_{Kt} - u_{Kxx} - f(u_K) \leq 0$, so u_K is a subsolution of the original equation (4.1), and hence $u_K \leq u$. But by the result in part (a), $\lim_{t\to\infty} u_K(x,t) = \text{const} > K$, so that u itself eventually stays greater than K. Since K was arbitrarily large, $u \to \infty$. This completes the proof.

The analog of Theorem 4.8 holds for higher dimensional spaces, and for problems in domains with boundary, boundary conditions being imposed. However in the latter cases, w is not necessarily constant.

THEOREM 4.9. Let ψ be a bounded solution of (4.6) with a maximum or minimum at a finite value of x. Then it is unstable in the C^0 sense.

In fact, we shall show that if u is the solution of (4.1) with initial data $u(x,0) = u_o(x)$ such that $u_o(x) - \psi(x) \equiv v(x) \geq 0$, $v(x) \neq 0$, then $\sup |u(x,t) - \psi(x)| \geq \eta > 0$ for large enough t, where η does not depend on u_o.

We give the proof only for the case of ψ periodic. The proof for the other case, when ψ has a single maximum or minimum, is slightly more complicated, but follows the same lines.

Let E be the "energy" of ψ and m the maximum (see figure).

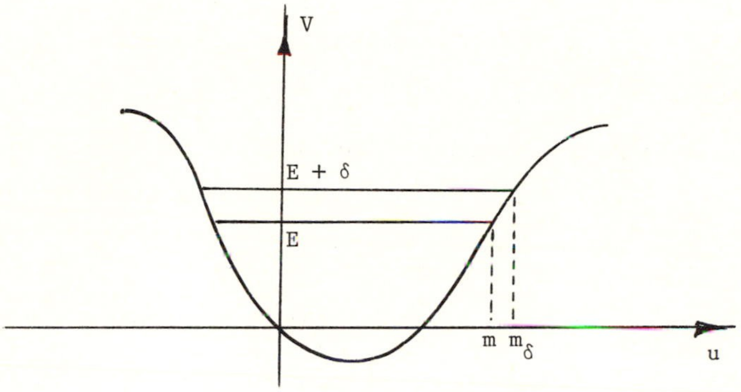

Figure 4.3.

Since $V'(m) > 0$, we know that for small enough $\delta > 0$, the solution corresponding to the segment with ordinate $E + \delta$ overlying the original segment has finite maximum $m_\delta > m$, and $\lim_{\delta \downarrow 0} m_\delta = m$. Shift the variable x so that the maxima of ψ and ψ_δ are attained at $x = 0$. Let I_δ be the interval containing the origin on which $\psi_\delta \geq \psi$. Since ψ_δ attains values less than $\min \psi$, we deduce that I_δ is bounded independently of δ. Let

$$\underline{\phi}(x) \equiv \begin{cases} \psi_\delta(x), & x \in I_\delta \\ \psi(x), & x \notin I_\delta . \end{cases}$$

Being locally the maximum of two solutions, $\underline{\phi}$ is a subsolution; and by continuous dependence on initial data,

$$\lim_{\delta \downarrow 0} |\underline{\phi} - \psi|_o = 0. \tag{4.9}$$

Let v be a solution of (4.1) with $v(x,0) = \underline{\phi}(x)$. By Theorem 4.8, either $v \to \infty$ or $v \to w(x)$, the least time-independent solution $\geq \underline{\phi}$. But it is evident from the characterization of such solutions in terms of line segments and zeros of f that w is a constant $> m$. Therefore no matter how small δ is, v will leave a fixed neighborhood of ψ, and ψ is therefore unstable. The same is true of any solution $u \geq v$.

Now let $\phi(x) \geq \psi(x)$, and $\phi \not\equiv \psi$. Let ϕ be the initial value of a solution u. By Theorem 4.1, $u > \psi$ for $t > 0$. Therefore since I_δ is bounded, we know that for small enough δ the subsolution $\underline{\phi}$ constructed above will satisfy $\underline{\phi}(x) < u(x,1)$. It follows that $u(x,t + 1) > v(x,t)$, so that u also leaves a neighborhood of ϕ, even though $u(x,0)$ differs from ψ by an arbitrarily small amount over an arbitrarily small interval. Aronson and Weinberger (1975) call this type of instability the "hair-trigger" effect. This same hair-trigger effect holds for unstable constant solutions.

The only bounded stationary solutions of (4.6) are those with maximum or minimum at a finite value of x, the monotone ones, and the constant ones. We have shown above that solutions in the first category are unstable. It will be shown later that those in the second are stable. For completeness we now consider the constant solutions.

THEOREM 4.10. Let u_o be a zero of f. It is C^o-stable as a solution of (4.1) if and only if it is stable as a solution of

$$\frac{du}{dt} = f(u). \tag{4.10}$$

(For the latter equation, stability is an easy matter to resolve.)

<u>Proof</u>: Suppose u_o is a stable state of (4.10). Given $\varepsilon > 0$, there exists $\delta > 0$ such that the solutions $u_\pm(t)$ of (4.10) satisfying $u_\pm(0) = u_o \pm \delta$ also satisfy $|u_\pm(t) - u_o| < \varepsilon$. But $u_\pm$ are also solutions of (4.1), so by Theorem 4.1, if $u(x,t)$ is a solution of (4.1) satisfying $|u(x,0) - u_o| < \delta$, then

necessarily u exists for all time, and $u_- \le u \le u_+$, so $|u(x,t) - u_0| < \varepsilon$, $t > 0$.

Now suppose u_0 is unstable for (4.10). Then for some ε, every solution of (4.10) except $u \equiv u_0$ leaves the ε-neighborhood of u_0. But solutions of (4.10) are also solutions of (4.1), so u_0 is C^0-unstable for (4.1) as well.

As was mentioned before, the monotone nonconstant solutions are stable. However, such solutions exist only in the special case that there is a line segment joining two neighboring maxima of exactly the same height (see (d) in the Figure 4.2). Usually there will be no such pair of adjacent maxima, and when there is a slight change in the function f will destroy this property. This type of solution, therefore, though stable, is structurally unstable since it is destroyed by a small change in the equation itself (the way it is destroyed is that it is converted into a front with nonzero velocity).

To summarize the above results, we have the following statement: The only bounded stable stationary solutions of (4.1) are (1) the constants which are stable as rest states of (4.10), and (2) the monotone ones; and the latter are structurally unstable.

Now let us set about estimating the domain of attraction of stable constant solutions. For illustration we take the case when f has two stable zeros, and suppose them to be 0 and 1. Thus $f(0) = f(1) = 0$, $f_u(0) < 0$, $f_u(1) < 0$, and the potential function $V = V(u)$ has two local maxima at 0 and 1. Let us assume $V(0) > V(1)$, and that there are no other maxima in $(0,1)$ which are as great as $V(1)$. In other words,

$$\int_0^u f(s)ds < 0 \quad \text{for} \quad u \in (0,1].$$

(See Figure 4.4.)

<u>Definition</u>: If $u_0 < u_1$ are stable constant states of (4.1), then u_0 is said to "dominate" u_1 if $\int_{u_0}^{u} f(s)ds < 0$ for all $u \neq u_0$ in the closed set between u_0 and u_1.

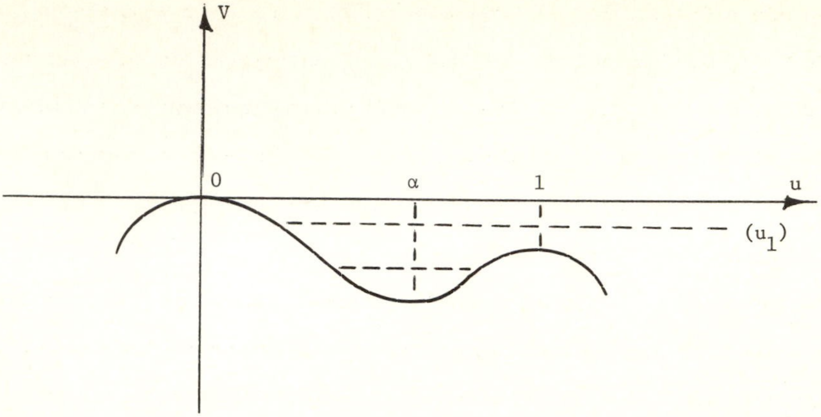

Figure 4.4.

Thus in the present example, 0 dominates 1, and the reverse would be true if $\int_u^1 f(s)ds = -\int_1^u f(s)ds > 0$ for $u \in [0,1)$.

Let $u_1(x)$ be a solution represented by the top dotted line in the figure, with E satisfying $V(1) < E < V(0)$. It is unbounded above, and has a single finite minimum u_m (which depends on E). Translate x so that this minimum is assumed at the origin.

PROPOSITION 4.11 (Aronson and Weinberger 1975). Let $\phi(x)$ satisfy $0 \leq \phi \leq 1$, and $\phi(x) \leq u_1(x)$ for some u_1 as described. Then the solution $u(x,t)$ with initial data $u(x,0) = \phi(x)$ satisfies $\lim_{t\to\infty} u(x,t) = 0$, uniformly on bounded intervals on the x-axis.

<u>Proof</u>: Let $\bar{\phi}(x) = \text{Min}\,[1, u_1(x)]$ (as in the figure).

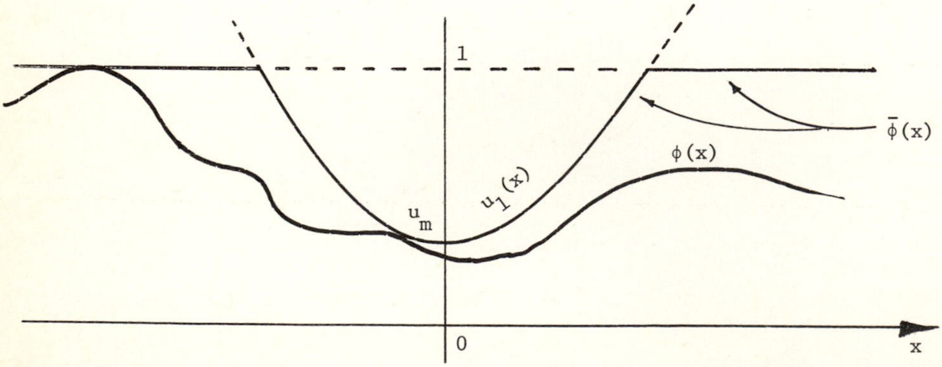

Figure 4.5.

It is a subsolution, and 0 is the least constant solution less than $\bar{\phi}$. It therefore follows from Theorem 4.8 (rather the analogous theorem using a supersolution) that $u \to 0$ as $t \to \infty$. The uniformity of the convergence on bounded intervals follows from the fact (Theorem 4.5) that u_x is uniformly bounded in t. In fact, this makes the functions $u(\cdot,t)$ equicontinuous, so the result follows from Arzela's theorem.

Now let α be a zero of f in the interval $(0,1)$ (such a point clearly exists - take, for example, the minimum of V, as in Figure 4.4). Also let $u_2(x)$ be a periodic solution with α in its range. This might correspond to the lower horizontal dotted line in that same figure. Take $x = 0$ to be a minimum of u_2, and let I be the interval centered at zero (as in Figure 4.6) with exactly four points where $u_2 = \alpha$, two of them being the endpoints. Let

$$\underline{\phi}(x) = \begin{cases} u_2(x), & x \in I \\ \alpha, & x \notin I. \end{cases}$$

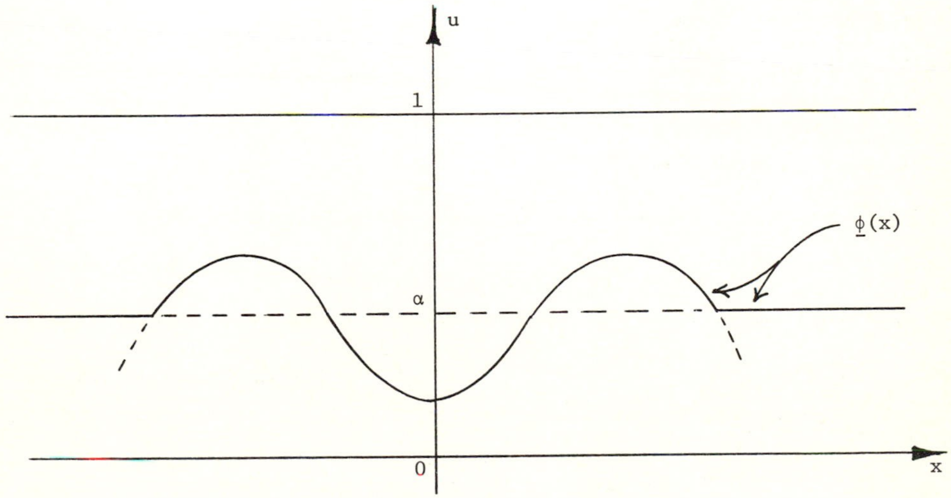

Figure 4.6.

Then $\underline{\phi}$, being locally the maximum of two solutions, is a subsolution.

PROPOSITION 4.12. Let ϕ satisfy $\underline{\phi}(x) \le \phi \le 1$ and u a solution with initial data $u(x,0) = \phi(x)$. Then $\lim_{t\to\infty} u(x,t) = 1$, uniformly on bounded intervals.

The proof is the same as before.

We have thus given two rather general conditions on ϕ, one of them sufficient for ϕ to be in the domain of attraction of 1, and the other sufficient for ϕ to be in the domain of attraction of 0. Note that the conditions are quite different, depending on whether the attractor in question is the dominant one (in this case, 0) or not. In fact, Proposition 4.11 shows that it is possible for ϕ to be in the domain of attraction of 0, the dominant state, even though $\phi = 1$ except for a bounded interval. This same proposition, however, shows that ϕ must be in the domain of attraction of 0 if $0 \le \phi \le 1$ and $\phi = 0$ for a sufficiently large (but bounded!) interval.

These are really threshold results, as is the following, even more clearly so.

PROPOSITION 4.13. Assume (as in Figure 4.4) that there is only one zero α of f for $u \in (0,1)$. Assume $\int_0^1 f(s)ds < 0$. Let $0 \le \phi \le 1$ and $\phi(x) \le \alpha - \eta$ for $|x| < L$, where η is any positive number, and L is sufficiently large, depending on η. Then $\lim_{t\to\infty} u(x,t) = 0$, uniformly on bounded sets.

The proof will not be given here (see Fife and McLeod 1977).

Thus, the first proposition requires ϕ to be sufficiently near 0 for a sufficiently long interval, whereas this last proposition requires ϕ to be less than α for a (perhaps longer) sufficiently long interval.

The methods of this section apply also to establish the instability of nonconstant solutions on bounded intervals satisfying zero Neumann conditions. Casten and Holland (1978) have proved instability results for nonconstant solutions in more than one space variable, on bounded domains, again satisfying zero Neumann conditions. The stability question for constant solutions with respect to perturbations which decay as $|x| \to \infty$ was examined by Chafee (1974). Chafee (1976) also considered the concepts of stable and unstable manifolds in connection with stationary solutions of (4.1). See Henry (1978) for a thorough treatment of this and many other questions of a geometric nature.

Apart from questions of stability, steady solutions of (4.1) were studied as a limiting case of model chemical schemes, in Boa (1975) and Ibañez and Velarde (1977).

4.4 Traveling waves

Here we consider the existence of traveling wave solutions of (4.1); that is, solutions of the form

$$u(x,t) = U(x - ct) \tag{4.11}$$

for some constant c. For the scalar equation (4.1), the only stable nontrivial traveling waves which exist are among those of wave front type, i.e., those for which $U(z)$ approaches distinct limits as $z \to \pm\infty$. For such to exist, it is clearly necessary that these distinct limits be zeros of f, since u_t and u_{xx} will vanish as $x \to \pm\infty$. Conversely, if f has two distinct zeros, u_o and u_1, then typically (but not always) a wave front does exist such that $U(-\infty) = u_o$ and $U(\infty) = u_1$. In particular, this is true if f has no zeros between the two given ones, or if it has only one zero between the two, at which $f' > 0$. When one of the u_i is an unstable rest point of the ordinary differential equation (4.10), then the traveling wave will be unstable for (4.1) in the C^o norm, but nevertheless may be significant in some applied context (as in Fisher's model (1937)).

In this section, we indicate how the existence of wave fronts may be obtained in the important cases, and develop some of their basic properties. In the following section, we estimate the domains of attraction of these fronts.

We assume that $f(0) = f(1) = 0$ (and that $f \in C^1[0,1]$, as always), and look for traveling fronts with range $0 \leq U \leq 1$.

The first point to be made is that such fronts are necessarily monotone. In fact, any solution $u = U(x - ct)$ of (4.1) in this range, with $U(-\infty) = 0$, $U(\infty) = 1$, necessarily satisfies $U'(z) > 0$ for finite $z = x - ct$.

To see this, we first substitute the expression (4.11) into (4.1) to obtain

$$U'' + cU' + f(U) = 0. \tag{4.12}$$

Therefore the front corresponds to a trajectory, in the (U,P) phase plane, of the system

$$\frac{dU}{dz} = P, \tag{4.13a}$$

$$\frac{dP}{dz} = -cP - f(u), \tag{4.13b}$$

connecting the stationary points $(0,0)$ and $(1,0)$. This trajectory stays in the strip $0 \leq U \leq 1$, and has the property that it is directed toward the right for $P > 0$, and toward the left for $P < 0$. Any simple curve with these properties must be such that $P \geq 0$ throughout its length. If it contains a point $(U_o,0)$ with $U_o \in (0,1)$, then there would exist a traveling front $U(x)$ such that $U(0) = U_o$, $U'(0) = 0$. Then $U''(0) \neq 0$, for otherwise by uniqueness of solutions of (4.12), $U \equiv U_o$. This means that P would change sign as the point $(U_o,0)$ is crossed, which we have seen to be impossible. Therefore $P = U' > 0$ except at the endpoints, which is what we wanted to show.

In view of this result, to any traveling front with range $[0,1]$ there corresponds a function $P(U)$ defined for $U \in [0,1]$, positive in $(0,1)$, zero at $U = 0$ or 1, representing the derivative dU/dz. From (4.12), we see that it satisfies the equation

$$P' + \frac{f}{P} = -c, \tag{4.14}$$

where c is the corresponding wave speed. Moreover, P has to satisfy the boundary conditions

$$P(0) = P(1) = 0. \tag{4.15}$$

Conversely, given such a function P satisfying (4.14-4.15), we may obtain a corresponding solution of (4.12) by integrating

$$U'(z) = P(U), \qquad U(0) = \frac{1}{2}.$$

This equation may be solved for z in an interval (z_0, z_1) to obtain a monotone solution with $\lim_{z \downarrow z_0} U(z) = 0$, $\lim_{z \uparrow z_1} U(z) = 1$. To show that $u(x,t) = U(x - ct)$ is a traveling front as we have defined it, we have only to verify that $z_0 = -\infty$, $z_1 = \infty$.

Since $f(0) = 0$, we have that $|f(U)| < \beta U$ for some β. Let γ be a positive number such that $\frac{\beta}{\gamma} - c < \gamma$. Let S be the line $P = \gamma U$ in the (U,P) plane. If the graph of the given solution $P(U)$ touches S at a point in the first quadrant distinct from the origin, then at that point, $P' = -c - \frac{f}{P} \leq -c + \frac{\beta}{\gamma} < \gamma$, so that the trajectory immediately goes below S. This implies that for some $\delta > 0$, either

(i) $P(U) > \gamma U$ for $U \in (0,\delta)$, or

(ii) $P(U) < \gamma U$ for $U \in (0,\delta)$.

In the former case we have, from (4.14),

$$P'(U) = -c - \frac{f}{P} \leq -c + \frac{\beta}{\gamma} < \gamma, \qquad \text{so that} \qquad P(U) \leq \gamma U.$$

Therefore (ii) must hold. But then

$$-z_0 = \int_0^{1/2} \frac{du}{P(u)} > \frac{1}{\gamma} \int_0^{1/2} \frac{du}{u} = \infty, \qquad \text{so that} \qquad z_0 = -\infty.$$

Similarly, $z_1 = \infty$.

Hence if $f \in C^1[0,1]$, $f(0) = f(1) = 0$, there is a one-one correspondence between traveling fronts (modulo shifts in the independent variable z) and solutions of (4.14-4.15), positive in $(0,1)$.

The form of the equation (4.14) makes it clear that for every solution pair (P,c), there is a second pair $(-P,-c)$, so our theory applies to monotone decreasing solutions of (4.12) as well.

Integration of (4.14) (after multiplication by P) yields

$$c\int_0^1 P(u)du = -\int_0^1 f(u)du,$$

so that, for a positive solution of (4.14-4.15), we have

$$c \gtreqless 0 \quad \text{according as} \quad \int_0^1 f(u)du \lesseqgtr 0. \tag{4.16}$$

(For a negative solution, the sign of c is the same as that of $\int$ fdu.)

To find a traveling front, we have to find, for some c, an orbit of the system (4.13) connecting the critical points (0,0) and (1,0). The existence and uniqueness of such an orbit depends strongly on the types of critical points they are. To determine the types, we linearize (4.13) about the points in question. First we consider the point (0,0). The linearization at this point is the system

$$\frac{dW}{dz} = AW,$$

where $W = (U,P)$ and $A = \begin{pmatrix} 0 & 1 \\ -f'(0) & -c \end{pmatrix}$. The eigenvalues of A are given by

$$\lambda = \frac{-c \pm \sqrt{c^2 - 4f'(0)}}{2}. \tag{4.17}$$

There are three important cases to consider, and they are listed below. The only remaining case is the borderline one when $f'(0) = 0$, and we ignore it.

Case 1: $f'(0) > 0$, $c^2 \geq 4f'(0)$. In this case, both eigenvalues are real and of the same sign, so the critical point (0,0) is a "node," and all orbits near it either leave it or enter it. By (4.13a), orbits in the first quadrant have U increasing, so in fact all orbits near (0,0) must emanate from that point, as shown:

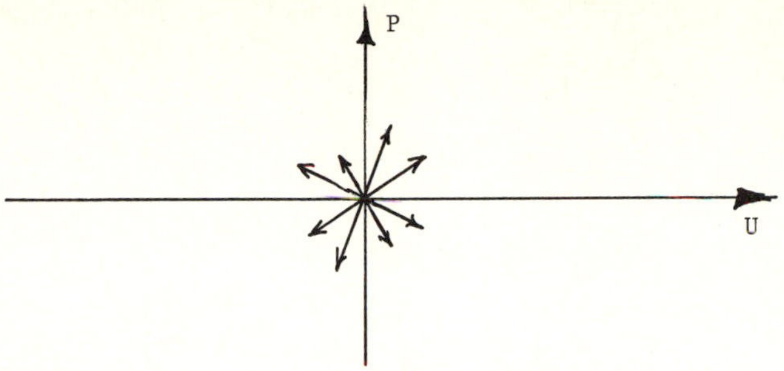

This means that the λ's are positive, so from (4.17), Case 1 can only happen when $c < 0$.

Case 2: $f'(0) < 0$. In this case, the eigenvalues are real and of opposite sign, so the critical point is a "saddle" as shown (again, consider (4.13a)):

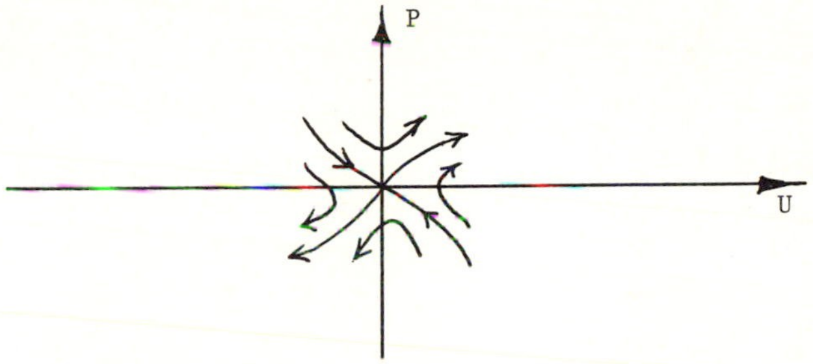

Case 3: $c^2 < 4f'(0)$. In this case, the eigenvalues are complex, and (0,0) is a spiral or a center, as shown:

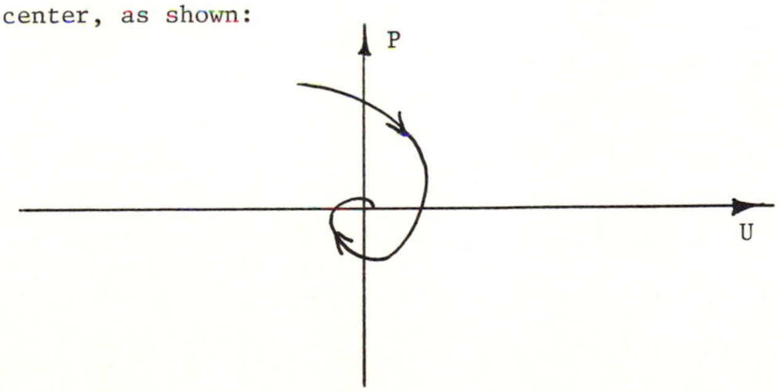

Case 3 clearly excludes fronts with range [0,1], so we concentrate on Cases 1 and 2. A similar categorization may be made at the other point (1,0), which could be a node or a saddle. This makes four possible combinations in all, but on further thought, some may be ruled out.

Thus if (0,0) and (1,0) were both nodes, then as we have seen, (0,0) must be an "unstable" node (orbits come out of (0,0)). But the same argument says that the opposite must be true of (1,0). The former requirement implies that

$c < 0$, whereas the latter forces $c > 0$. This case is therefore impossible.

If $(0,0)$ is a saddle $(f'(0) < 0)$ and $(1,0)$ a node $(f'(1) > 0)$, the simple transformation $\hat{U} = 1 - U$, $\hat{f}(\hat{U}) = -f(1 - \hat{U})$ reduces the problem to one for which the opposite is true: $(0,0)$ a node and $(1,0)$ a saddle.

We therefore need consider only two cases: (a) a saddle-saddle transition, and (b) a node-saddle transition. In both cases, $f'(1) < 0$ (since $(1,0)$ is a saddle), and in case (a), when $(0,0)$ is a node, we restrict c so that $c^2 \geq 4f'(0)$.

(a) Saddle-saddle orbits: Here we assume $f'(0) < 0$, $f'(1) < 0$, and look for a solution P of (4.14-4.15) with $P(0) > 0$ for $U \in (0,1)$. It turns out that under general assumptions, there is a unique value of c for which such a solution exists. Our analysis uses the following monotonicity lemma.

LEMMA 4.14 (Kanel' 1962). Let $f(u) \leq 0$ for small positive u. Let $P_i(U)$ $(i = 1, 2)$ be two solutions of (4.14), $P_i(0) = 0$, entering the positive quadrant from the origin, with two corresponding values of $c = c_i$. If $c_1 = c_2$, then $P_1(U) \equiv P_2(U)$ wherever $P_1(U) > 0$, $U > 0$. If $c_1 > c_2$, then $P_1(U) < P_2(U)$ wherever $P_1(U) > 0$, $U > 0$.

Proof: From (4.14), we have

$$P_1' - P_2' - \frac{f}{P_1P_2}(P_1 - P_2) = -(c_1 - c_2),$$

so that

$$\frac{dG(U)}{dU} = -(c_1 - c_2) \exp \int_{U_0/2}^{U} (-f(t)/P_1(t)P_2(t))dt,$$

where

$$G(U) \equiv (P_1 - P_2) \exp \int_{U_0/2}^{U} (-f(t)/P_1(t)P_2(t))dt.$$

As $U \downarrow 0$, we have $G(U) \to 0$ since $P_1 - P_2 \to 0$ and the exponential factor is bounded as $U \downarrow 0$ because of the sign of f. If $c_1 = c_2$, $G(U)$, being constant, is zero, so that $P_1 \equiv P_2$. But if $c_1 > c_2$, G is strictly decreasing, so that $P_1 < P_2$ for $U > 0$.

Thus, increasing c decreases P. It is clear that the requirement $f < 0$ for small u is necessary for the validity of this result, because in the node case there are many trajectories leaving the origin.

The lemma indicates that for each c, there is at most one trajectory entering the first quadrant from the origin. This is also evident from the fact that the two eigenvalues are of opposite signs. This latter fact also tells us that there does exist such a trajectory.

Let T_c be this trajectory. It is bounded, because from (4.14), P' is bounded for P bounded away from zero. It must therefore leave the half-strip $Q = \{U \in (0,1),\ P > 0\}$ first at some point (U_c, P_c), where either $0 < U_c \leq 1$ and $P_c = 0$, or $U_c = 1$ and $P_c > 0$. This is because by (4.13a), U is increasing for $P > 0$.

Suppose $c < 0$ and $-c$ is large. If, on T_c, P attains a maximum P_{max} for $U < 1$, then at that point $P' = -c - \frac{f}{P} = 0$, so $P_{max} = -\frac{f}{c} = \frac{f}{|c|} > 0$. Therefore that maximum must occur for $U > U_o$, the first zero of f in $(0,1)$. But for $U < U_o$, $f(U) < 0$, so $P'(U) = -c - \frac{f}{P} \geq -c = |c|$, and so $P(U_o) \geq |c|U_o$. Therefore $|c|U_o \leq P_{max} = \frac{f(U_{max})}{|c|}$, which is impossible for large $|c|$, and we conclude that for large negative c, P is strictly increasing on $(0,1)$, and $Q_c = 1$, $P_c > 0$.

When c is increased from such a value, $P(U)$ strictly decreases, by the lemma above, and does so continuously in c (we shall not prove this; see Aronson and Weinberger (1978)). There must eventually be a value of $c = c_o$ at which $P_{c_o} = 0$. And either $U_{c_o} = 1$ or $U_{c_o} < 1$, the latter case occurring when the trajectory with $c = c_o$ is the limit, as $c \uparrow c_o$, of trajectories with $P > 0$ on $(0,1)$ (as in the figure on the next page).

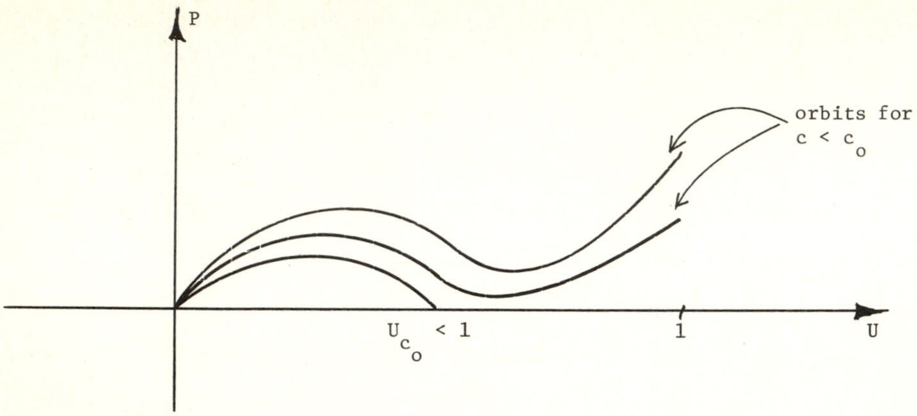

In the latter case, for c slightly less than c_o, P' will have at least two zeros U_1 and $U_2 \in (0,1)$:

$$0 = P' = -c - \frac{f(U_i)}{P(U_i)}, \quad i = 1, 2,$$

with $U_1 < U_2$, $P(U_1) > P(U_2)$, $\lim_{c \uparrow c_o} U_2 = U_{c_o}$, $\lim_{c \uparrow c_o} P(U_2) = 0$. But $-c = \frac{f(U_1)}{P(U_1)} = \frac{f(U_2)}{P(U_2)}$, so $f(U_1) > f(U_2)$. Moreover, by passing to the limit as $c \uparrow c_o$, we obtain $f(U_{c_o}) = 0$. Thus U_{c_o} is a zero of f. But it is not the first such zero in $(0,1)$, for $f(U) < 0$ for small positive U, and $f(U_1) > 0$, so there must be a zero in $(0,U_1)$ as well.

To sum up the above argument, we have found that the number of interior zeros of f is an important consideration. The case we are treating is when $f'(0) < 0$, $f'(1) < 0$, so there is at least one such interior zero. If we assume there is no more than one, then the second case $U_{c_o} < 1$, as described above, cannot occur, and we conclude that $U_{c_o} = 1$, $P_{c_o} = 0$, so that for $c = c_o$, there is a solution (by the lemma, unique) of (4.14-4.15). This does not tell us the value of c, although one can infer its sign from (4.16).

Our result is that there does exist a traveling front solution if f has only one interior zero. This is not a necessary condition, however; see Fife and McLeod

(1977) for a condition that is both sufficient and necessary, in the saddle-saddle case.

(b) Node-saddle orbits: In this case, the analysis and the results, are quite different. We assume now that $f'(0) > 0$, $f'(1) < 0$. To make things simpler, we also assume $f(u) > 0$ for $u \in (0,1)$. This includes the case $f(u) = u(1 - u)$ examined by Fisher (1937).

First, note that by (4.16), $c < 0$.

The critical point $(1,0)$ is a saddle, so by an argument similar to the one above, there exists a unique trajectory T "hitting" $(1,0)$ from the left and above, as in the figure below.

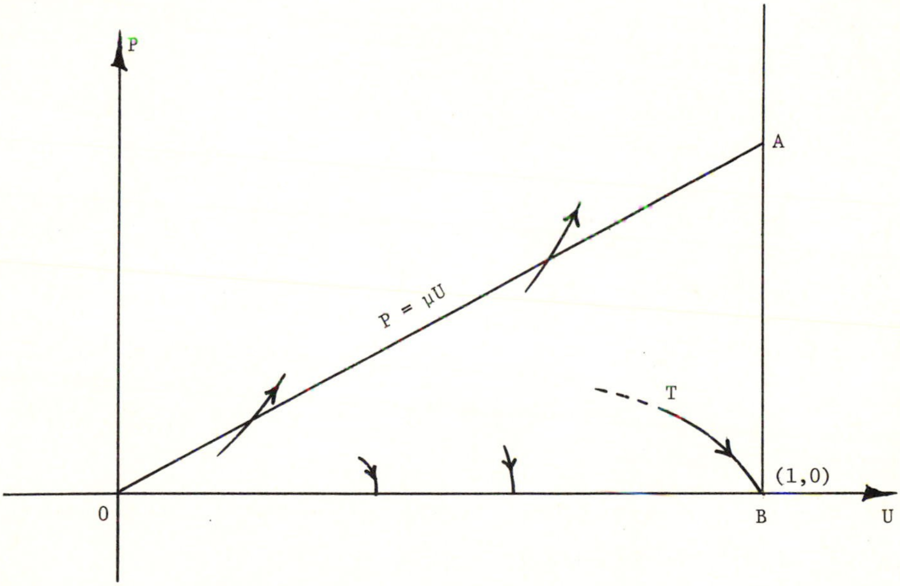

For some $\mu > 0$, consider the line $P = \mu U$ as indicated. Also label the points A and B as shown. We follow the trajectory T back into the triangle OBA, and ask where it eventually leads. It must either lead to a critical point in the closed triangle, or else cross the boundary. We consider the various possibilities in turn.

First, it easily checked that the only critical points in the closed triangle are $(0,0)$ and $(1,0)$.

Next, consider the possibility of T crossing the boundary. Recall that by

(4.13a), U is strictly increasing on T. Therefore T cannot attain the side AB.

For $P > 0$ but small and $U \in (0,1)$,

$$\frac{dU}{dP} = \frac{P}{-cP - f(U)} < 0,$$

so the trajectories approaching the segment OB do so from the left. Therefore T cannot reach this side, at any point with $U > 0$.

On the diagonal line OA,

$$\frac{dP}{dU} = -c - \frac{f(U)}{P} \geq -c - \frac{\nu U}{P},$$

where

$$\nu = \sup \frac{f(U)}{U} \geq f'(0).$$

Thus on OA, $\frac{dP}{dU} \geq -c - \frac{\nu}{\mu}$. If the positive number μ satisfies

$$-c - \frac{\nu}{\mu} \geq \mu, \tag{4.18}$$

then any trajectory leaving the triangle through OA must do so from the left (as shown in the previous figure), and this excludes T.

Therefore if (4.18) holds, the only possibility is for T to lead to the node at the origin, and we again have the existence of a wave front.

Now (4.18) can be written as

$$\mu^2 + c\mu + \nu \leq 0.$$

This inequality is equivalent to the statements that

(1) the two roots $(-c \pm \sqrt{c^2 - 4\nu})/2$ are real, and

(2) μ lies between them.

Let us assume that c satisfies the inequality

$$c \leq -\sqrt{4\nu} < 0. \tag{4.19}$$

Then (1) holds, and there exists a positive value of μ satisfying (2). Furthermore, our original hypothesis $c^2 \geq 4f'(0)$ will not be violated, because $\nu \geq f'(0)$.

Therefore for <u>every</u> value of c satisfying (4.19), there exists a wave front orbit connecting $(0,0)$ to $(1,0)$. Actually, this is still not a necessary condition. The fact is, there exists a maximal speed c^* somewhere in the interval

$$-\sqrt{4\nu} \leq c^* \leq -\sqrt{4f'(0)},$$

such that for every $c \leq c^*$, there exists a wave front.

The following theorem recapitulates the above results:

THEOREM 4.15. Assume $f(0) = f(1) = 0$.

(a) If $f'(0) < 0$, $f'(1) < 0$, and f has only one zero in $(0,1)$, then there exists a unique (except for translation) wave front solution of (4.1) with $U(-\infty) = 0$, $U(\infty) = 1$.

(b) If $f'(0) > 0$, $f'(1) < 0$, and $f(u) > 0$ for $u \in (0,1)$, then there exists a number $c^* < 0$ such that there exists a wave front solution with $U(-\infty) = 0$, $U(\infty) = 1$ if and only if $c \leq c^*$.

The case (b) with $f'(u) \leq f'(0)$ was studied in the pioneering work of Kolmogorov, Petrovskiĭ, and Piscunov (1937). They proved the existence of the family of wave fronts and moreover proved that a solution of (4.1) which starts at $t = 0$ as a step function, equal to zero for $x < 0$ and to one for $f > 0$, converges in "wave form" to the front with speed c^* (we shall explain this type of convergence later). Kanel' (1962) proved the existence of a unique such front in the case, more general than (a) above, $f(u) \leq 0$ for $u \in (0,u_0)$, $f(u) > 0$ for $u \in (u_0,1)$. He also established certain convergence results. The case when $f \equiv 0$ for small positive u arises in the theory of combustion.

See Aronson and Weinberger (1975 and 1978) for a treatment of existence and other questions. See also Hadeler (1976) and Hadeler and Rothe (1974). Waves of type (b) exist for a wide class of 2-component RD systems (Tang and Fife 1979).

4.5 Global stability of traveling waves

Here we tackle the following question: What conditions on the initial data $u(x,0) = \phi(x)$ can one impose in order to guarantee that the solution $u(x,t)$ of (4.1) converges to a traveling front?

The first result of this type was that of Kolmogorov, Petrovskiǐ, and Piscunov (1937), who treated the case

$$f(u) > 0, \quad f'(u) < f'(0) \quad \text{for} \quad u \in (0,1); \quad f'(0) > 0,$$

with ϕ the step function

$$\phi(x) = \begin{cases} 0 & \text{for} \quad x < 0 \\ 1 & \text{for} \quad x > 0. \end{cases}$$

According to Theorem 4.15, there is a maximal wave velocity $c^* < 0$, with associated wave front U. The result of Kolmogorov, Petrovskiǐ, and Piscunov is that there exists a function ψ such that

$$|u(x,t) - U(x - c^*t - \psi(t))| \to 0 \quad \text{as} \quad t \to \infty,$$

uniformly in x: and $\lim_{t\to\infty} \psi'(t) = 0$.

Since it is not true that $\psi(t) \to 0$, the convergence of u to a traveling front is not uniform. Instead, we have convergence of "wave form" or "wave profile."

For other convergence results with this type of f, see Rothe (1975, 1978), Stokes (1976), Kametaka (1976), Hoppensteadt (1975), McKean (1975), Moet (1978),

Uchiyama (1977), and Bramson (1977). The last two authors obtained the asymptotic behavior of the function ψ. Kanel' (1962, 1964) gave some convergence results for functions f arising in combustion theory. A generalization to waves traveling along a strip was given by Larson (1977b).

Our attention in this section will be paid mainly to the case $f'(0) < 0$, $f'(1) < 0$, as in Theorem 4.15(a). This case arises not only in selection-migration models (Chapter 2), but in other simple population models with two stable states, and as a degenerate case of nerve signal equations. See the connection with transmission line-theory in Nagumo, Arimoto, and Yashizawa (1962).

In this case, one can prove uniform convergence to the wave front (with ψ = const) for a wide class of initial data (Fife and McLeod 1977, McLeod and Fife 1979). Convergence of wave form for cases in which f can also depend on u_x was established by Chueh (1975).

For simplicity, we assume that f has only one zero in $(0,1)$, at the point α. So we assume

$$f'(0) < 0, \quad f'(1) < 0, \quad f(u) < 0 \text{ for } u \in (0,\alpha), \quad f(u) > 0 \text{ for } u \in (\alpha,1). \tag{4.20}$$

THEOREM 4.16. If (4.20) holds, and if ϕ satisfies

$$\limsup_{x\to-\infty} \phi(x) < \alpha, \qquad \liminf_{x\to\infty} \phi(x) > \alpha, \tag{4.21}$$

then for some constants z_o, C, and w, the last two positive,

$$|u(x,t) - U(x - ct - z_o)| < Ke^{-wt}.$$

We use a moving coordinate system z, t, where $z = x - ct$, and write $v(z,t) = u(x,t) = u(z + ct,t)$. Then (4.1) becomes

$$N[v] \equiv v_t - cv_z - v_{zz} - f(v) = 0. \tag{4.22}$$

The principal ingredient in the proof of Theorem 4.16 is the following:

LEMMA 4.17. Under the hypotheses of Theorem 4.16, there are constants z_1, z_2, q_0, and μ, the last two positive, such that

$$U(z - z_1) - q_0 e^{-\mu t} \leq v(z,t) \leq U(z - z_1) + q_0 e^{-\mu t}. \tag{4.23}$$

Proof: We prove only the left-hand inequality; the other is similar. The function v satisfies (4.22). Functions $\xi(t)$ and $q(t)$ (q positive) will be chosen so that

$$\underline{v}(z,t) \equiv \text{Max } [0, U(z - \xi(t)) - q(t)]$$

will be a subsolution.

First, let $q_0 > 0$ be any number such that $\alpha < 1 - q_0 < \liminf_{z\to\infty} \phi(z)$. Then take z^* so that $U(z - z^*) - q_0 \leq \phi(z)$ for all z. This is possible, for sufficiently large positive z^*, by virtue of (4.21). This situation is shown in the following figure.

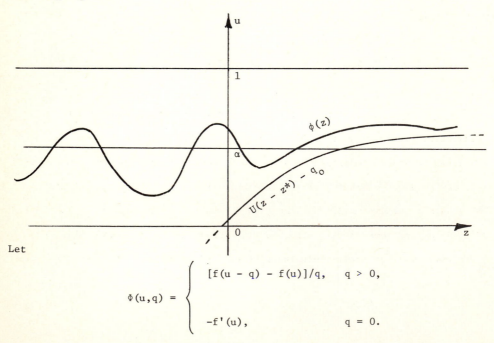

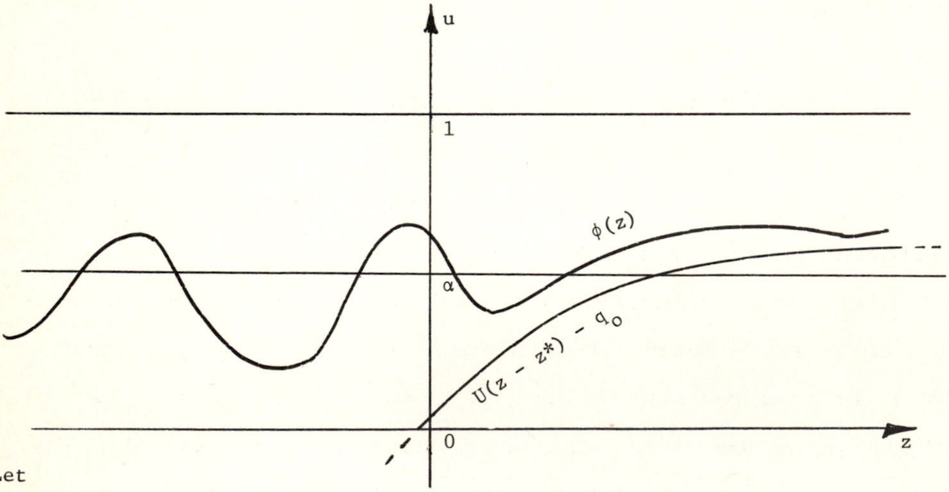

Let

$$\Phi(u,q) = \begin{cases} [f(u - q) - f(u)]/q, & q > 0, \\ -f'(u), & q = 0. \end{cases}$$

Then Φ is continuous for $q \geq 0$, and for $0 < q \leq q_o$ we have $\alpha_1 < 1 - q_o \leq 1 - q < 1$, so that $\Phi(1,q) > 0$. Also $\Phi(1,0) = -f'(1) > 0$. Thus for some $\mu > 0$, we have $\Phi(1,q) \geq 2\mu$ for $0 \leq q \leq q_o$. By continuity, there exists a $\delta > 0$ such that $\Phi(u,q) \geq \mu$ for $1 - \delta \leq u \leq 1$, $0 \leq q \leq q_o$. In this range, we have

$$f(u - q) - f(u) \geq \mu q.$$

Setting $\zeta = z - \xi(t)$, and using the fact that (4.12)

$$U'' + cU' + f(U) = 0, \tag{4.24}$$

we find that, if $\underline{v} > 0$,

$$N[\underline{v}] = -\xi'(t)U'(\zeta) - cU'(\zeta) - q'(t) - U''(\zeta) - f(U - q)$$

$$= -\xi'(t)U'(\zeta) - q'(t) + f(U) - f(U - q).$$

Thus when $U \in [1 - \delta,1]$, $q \in [0,q_o]$,

$$N[\underline{v}] \leq -\xi'U' - q' - \mu q \leq -(q' + \mu q),$$

provided $\xi' \geq 0$, since $U' \geq 0$ (see Section 4.4). We choose $q(t) = q_o e^{-\mu t}$, which results in $N[\underline{v}] \leq 0$ when $1 - \delta \leq U \leq 1$.

By possibly further reducing the size of μ and δ and using the same arguments, we may be assured that $N[\underline{v}] \leq 0$ whenever $0 \leq U \leq \delta$, $U \geq q$, as well.

Now consider the intermediate values, $\delta \leq U \leq 1 - \delta$. In this range, we know that $U'(z) \geq \beta$ for some $\beta > 0$. This fact was shown in Section 4.4. Also, by the differentiability of f, we have that $f(U) - f(u - q) \leq \kappa q$ for some $\kappa > 0$. Thus

$$N[\underline{v}] \leq -\beta\xi' - q' + \kappa q.$$

We now set

$$\xi'(t) = (-q' + \kappa q)/\beta = (\mu + \kappa)q/\beta > 0, \quad \text{with} \quad \xi(0) = z^*.$$

(Specifically,

$$\xi = z_1 + z_2 e^{-\mu t},$$

where

$$z_2 = -q_o(\mu + \kappa)/\mu\beta, \quad z_1 = z^* - z_2.)$$

Thus $\xi(t)$ is increasing and approaches a finite limit as $t \to \infty$. Then $N[\underline{v}] \leq 0$ whenever $\underline{v} > 0$, and by our condition on z^*, $\underline{v}$ will be a subsolution. Thus

$$v(z,t) \geq \underline{v}(z,t) \geq U(z - z_1) - q(t) = U(z - z_1) - q_o e^{-\mu t},$$

which completes the proof.

COROLLARY 4.18. Under assumption (4.20), the wave front $U(x - ct - z_o)$ is a stable solution of (4.1).

<u>Proof</u>: Given any $\varepsilon > 0$, we may guarantee that

$$|v(z,t) - U(z - z_o)| < \varepsilon$$

for all $t > 0$ by taking $|\phi(z) - U(z - z_o)|$ small enough. In fact, in the proof of the preceding lemma, we may take $q_o = O(\varepsilon)$, and $|z^* - z_o| = O(\varepsilon)$. Hence also $|z_1 - z_o| = O(\varepsilon)$, $|z_2 - z_o| = O(\varepsilon)$, and the conclusion follows from that of the lemma.

Although we have the stability of U already at this stage, Theorem 4.16 claims much more. We shall omit some of the details in the rest of the proof.

LEMMA 4.19. For each $\delta > 0$, the "orbit" set

$$\{v(\cdot,t): t \geq \delta\},$$

considered as a subset of $C^2(-\infty,\infty)$, is relatively compact.

Proof: Since $0 \leq \psi \leq 1$ and $v \equiv 0$, $v \equiv 1$ are solutions (hence subsolution and supersolution, respectively), we have that $0 \leq v(z,t) \leq 1$. Let $\{t_n\}$ be any sequence tending to ∞. According to Theorem 4.6, or rather to the analogous theorem for (4.22) in place of (4.1), there is a subsequence $\{t_n'\}$ such that $v(z,t_n')$ will converge to a limit $W(z)$, uniformly on bounded sets, and the same will be true of the x-derivatives up through order two. This does not establish convergence in C^2, because the uniformity is only on bounded sets. Nevertheless Lemma 4.17 shows that $v(z,t)$ is arbitrarily close to constants for large enough z and t. Theorem 4.5 then also shows that its derivatives are arbitrarily small for large z and t. This added information, controlling the behavior of v away from bounded intervals, is enough to establish that $v(z,t_n')$ converges to $W(z)$ uniformly in z, together with its derivatives. This is what we need for compactness, and finishes the proof of the lemma.

What we must show now is that (1) W is a wave front, and (2) the convergence of v to W is more than just along a sequence of t-values.

The first step in the proof will be to "truncate" v for large z and t. This is a purely technical device occasioned by the fact that unless truncated, v will not necessarily be in the domain of the Lyapunov function we shall use.

Let ε be a small positive number to be specified later, and define w as follows:

$$w(z,t) = v(z,t) \quad \text{for} \quad |z| \leq \varepsilon t,$$

$$w(z,t) = 0 \quad \text{for} \quad z \leq -\varepsilon t - 1,$$

$$w(z,t) = 1 \quad \text{for} \quad z \leq \varepsilon t + 1,$$

and w is a smooth function in the intermediate intervals, whose derivatives satisfy essentially the same bounds as do those of v.

We define the Lyapunov function

$$V[w] = \int_{-\infty}^{\infty} e^{cz}[\frac{1}{2} w_z^2 - \Phi(w) + H(z)\Phi(1)]dz,$$

where $H(z)$ is the Heaviside step function, and $\Phi(v) \equiv \int_0^v f(s)ds$. It clearly converges, as do the integrals below, because of the truncation.

The remainder of the proof of Theorem 4.16 will be divided into several steps.

(1) $V[w(z,t)]$ is bounded independently of t. The reason for this is that as $z \to \infty$, $v_z \to 0$, and the rate can be estimated by means of Theorem 4.5 and (4.23). The rate is exponential, faster than e^{cz}. Using this, one obtains similar estimates for the rate of decay of $\Phi(v) - \Phi(1)$, and finally for V. It is here that ε must be chosen small enough.

(2) $\frac{d}{dt} V[w] \equiv \dot{V}(t)$ exists, and

$$\dot{V}(t) = -\int_{-\infty}^{\infty} e^{cz}[w_{zz} + cw_z + f(w)]w_t dz.$$

This formula is seen formally by taking the time derivative inside the integral sign and integrating by parts. It is justified by expressing the derivative as limit of a difference quotient.

(3) $$\limsup_{x\to\infty} \dot{V}(t) = 0. \tag{4.25}$$

First, we define

$$Q[w] = \int_{-\infty}^{\infty} e^{cz}[w_{zz} + cw_z + f(w)]^2 dz,$$

so that

$$\dot{V}(t) + Q[w(\cdot,t)] = \int_{-\infty}^{\infty} e^{cz}[w_{zz} + cw_z + f(w)]N[w]dz,$$

where N is the operator in (4.22). Since $N[v] = 0$ and w is a truncation of v, we can estimate the rate at which $N[w]$ approaches zero as $|z|$ and t approach infinity. Together with the other estimates we have for the derivatives of w, and the above integral representation, we obtain

$$\lim_{t\to\infty} (\dot{V}(t) + Q[w(\cdot,t)]) = 0. \tag{4.26}$$

But since $Q \geq 0$, we have

$$\limsup_{t\to\infty} \dot{V}(t) \leq 0.$$

But if it were true that $\limsup \dot{V} < 0$, then it would follow that $V[w] \to -\infty$, contradicting the result in item (1) above. This proves (4.25).

(4) There exists a sequence $\{t_n\}$ approaching infinity, with

$$\lim_{n\to\infty} Q[w(\cdot,t_n)] = 0.$$

This follows from (4.25) and (4.26).

(5) There exists a sequence t_n' such that for some z_o,

$$v(z,t_n') \to U(z - z_o) \tag{4.27}$$

uniformly in z, together with second derivatives. To see this, take the sequence in item (4). By the compactness Lemma 4.19, it has a subsequence along which v, hence w, converges to a limit W in C^2. Using the result in item (4), we obtain, for any interval (a,b),

$$0 \leq \int_a^b e^{cz}[W'' + cW' + f(W)]^2 dz = \lim_{n\to\infty} \int_a^b e^{cz}[w_{zz} + cw_z + f(w)]^2_{t=t_n'} dz$$

$$\leq \lim_{n\to\infty} Q[w(\cdot,t_n')] = 0.$$

Since the integrands are nonnegative, they must vanish. Therefore W must satisfy

(4.12). Being the limit of v, it must also satisfy $W(-\infty) = 0$, $W(\infty) = 1$. Therefore it represents a wave front solution of (4.1), and by the uniqueness Lemma 4.14, it can only be a translate of the given front U. Thus

$$W(z) = U(z - z_o).$$

This establishes (4.27).

(6) $$\lim_{t\to\infty} v(z,t) = U(z - z_o), \quad \text{uniformly in } z.$$

In fact, we know such a limit takes place on a sequence of t-values, by item (5). But Corollary 4.18 says that once v comes close to $U(z - z_o)$, it remains close. This proves convergence.

(7) The convergence of v to $U(z - z_o)$ is, in fact, exponential in time, as indicated in the statement of Theorem 4.16. We shall not give the proof of this.

This finishes our treatment of the global stability properties of wave fronts.

4.6 More on Lyapunov methods

The foregoing proof made use of a Lyapunov function. Arguments of this type are widespread and important, so it may be worthwhile giving here an outline of the basic ingredients in Lyapunov stability theory, applied to dynamical systems on a Banach space. For RD systems in particular, see the treatment in Henry (1978). The following is taken from Hale (1969).

Let X be a Banach space, and S a subset of X.

Definition: A dynamical system on S is a continuous function $G(t,\phi)$: $\overline{R^+} \times S \to S$ such that

(i) $G(0,\phi) = \phi$,

(ii) $G(t + s,\phi) = G(t,G(s,\phi))$.

Definition: The orbit $\Gamma(\phi)$ of ϕ is the set

$$\Gamma = \{G(t,\phi): t \geq 0\}.$$

The ω-limit set $\omega(\phi)$ is the set of all $\psi \in S$ such that there exists a sequence $\{t_n\}$, $t_n \to \infty$, with $G(t_n,\phi) \to \psi$.

An <u>invariant set</u> $\xi \subset S$ is such that there exists a continuous function $\hat{G}$: $R \times \xi \to \xi$ satisfying (i), (ii), with $\hat{G}(t,\phi) = G(t,\phi)$ for $t \geq 0$. ($\hat{G}$ thus extends G to negative values of t.)

A Lyapunov function on an open set $T \subset S$ is a real valued function $V(\phi)$, defined and continuous on $\bar{T}$, which satisfies

$$\dot{V}(\phi) \leq 0 \quad \text{for all} \quad \phi \in T, \quad \text{where}$$

$$\dot{V}(\phi) \equiv \limsup_{h \downarrow 0} \frac{1}{h} (V(G(h,\phi)) - V(\phi)).$$

<u>Results</u>: If $\Gamma(\phi)$ is relatively compact, then $\omega(\phi)$ is nonempty, compact, and connected. If, furthermore, there is a Lyapunov function on a set $T \supset \Gamma(\phi)$, then $G(t,\phi) \to M$ as $t \to \infty$, where M is the largest invariant set in $\{\phi \in \bar{T}: \dot{V}(\phi) = 0\}$. If this set is discrete, then $G(t,\phi)$ approaches a limit as $t \to \infty$.

In applications, we want a function V on a set T which contains $\{\phi: V(\phi) < \rho\}$. Then if ϕ is some element with $V(\phi) < \rho$, we will be assured that the entire orbit $\Gamma(\phi)$ remains in T.

In the case discussed in the preceding Section 4.5, $X = C^2(-\infty,\infty)$; V was as defined; T consisted of functions for which the integral defining V converges. The condition $\dot{V}(\phi) = 0$ already implies ϕ is a traveling front, so that $G(t,\phi)$ exists for $t < 0$ as well, so that ϕ is already in an invariant set. Therefore M will be the set of traveling fronts; any translation of one is also one, so that M is not discrete; it is a one-dimensional continuum. Because of this fact, we needed an extra argument to prove that $G(t,\phi)$ approaches a limit. This extra argument was the stability result in Corollary 4.18.

A fair amount of attention has been given to the application of Lyapunov methods to reaction-diffusion systems. See, for example, Henry (1977, 1978),

Diekmann and Temme (1976), Chafee (1976), Chafee and Infante (1974), A. Hastings (1978), Murray (1975), de Mottoni and Tesei (1978), Rothe and de Mottoni (1978), and Maginu (1975).

4.7 Further results in the bistable case

Consider functions f satisfying (4.20), though some relaxation of these restrictions is possible. The global stability of wave fronts for (4.1) in this case was proved in Section 4.5; specifically, if

$$\liminf_{|x|\to\infty} |u(x,0) - \alpha| > 0 \tag{4.28}$$

and $u(x,0) - \alpha$ has opposite signs for large positive and large negative values of x, then $u(x,t)$ converges uniformly to a wave front. If (4.28) holds and $u(x,0) - \alpha \equiv z(x)$ has the same sign for large positive and negative values of x, then other possibilities can occur. The most interesting situation is when this sign is opposite that of $J \equiv \int_0^1 f(u)du$, for otherwise u will approach the constant 0 or 1 uniformly. Then if z is of one sign for all x, u again approaches 0 or 1 uniformly. But if (say) $z(x) \geq \delta > 0$ on an interval of length $\geq L(\delta)$, $J > 0$, and $z < 0$ for large $|x|$, then u uniformly approaches a diverging pair of wave fronts. In the terminology of Section 3.2, u is in the asymptotic state [W], which will therefore clearly also be globally stable (Fife and McLeod 1977). And in general if the asymptotic sign of z is opposite that of J, it can be shown (Fife 1979) that u converges to a constant, to W, or is not a uniformly stable solution.

A survey of theory and some applications of the bistable equation is found in Fife (1978a).

4.8 Stationary solutions for x-dependent source function

The equation

$$\frac{d^2u}{dx^2} + f(x,u) = 0$$

has been studied in connection with selection-migration models for an inhomogenous environment (see Chapter 2). It is natural to seek stable stationary solutions. In population dynamical contexts, these are called clines. Questions of existence, uniqueness, bifurcation, and stability of clines have been examined by Conley (1975), Fleming (1975), Fife and Peletier (1977), and others.

Addendum to Sec. 4.3.

Interesting results for the scalar nonlinear diffusion equation in a bounded domain with more than one space dimension have recently been obtained by H. Matano (Asympototic behavior and stability of solutions of semilinear diffusion equations, Publ. Res. Inst. Math. Sci. Kyoto, to appear). Of especial note in this paper is the demonstration that for certain types of domains, there exist stable nonconstant stationary solutions of $\Delta u + f(u) = 0$ satisfying homogeneous Neumann boundary conditions. As in Casten and Holland (1978), Matano also proved that for other domains, all nonconstant stationary solutions are unstable. Various other results concerning the time-asymptotic behavior of solutions are contained in the paper, as well as in H. Matano, 1978, Convergence of solutions of one-dimensional semilinear parabolic equations, J. Math. Kyoto Univ. 18, 221-227.

5. SYSTEMS: COMPARISON TECHNIQUES

In this and the next three chapters, we explore the possibility of extending to systems of reaction-diffusion equations some of the methods encountered in the previous chapter for the scalar equation. At the same time, certain other methods will be described, which were not discussed in that chapter. Our emphasis will be on the problem of obtaining existence of solutions of the types spoken of in Section 3.2, and of analyzing their stability characteristics.

5.1 Basic comparison theorems

Analogs for the comparison techniques which have been so successful in treating scalar equations exist for systems as well; but their range of applicability is smaller and the results they lead to are weaker.

Comparison theorems in the general category of the one below, with its corollaries, have been obtained by various people (Amann 1978; Hagan 1979; Conway and Smoller 1977,a,b,c; Chueh, Conley, and Smoller 1977; Weinberger 1975; Bebernes and Schmitt 1977; Chandra and Davis 1978; Auchmuty 1978; Nickel 1978). The same idea was used by Sattinger (1975) in connection with a problem in combustion theory. The linear case is found in Protter and Weinberger (1967), page 190. Hagan (1979) allows for time delays and quite general nonlinearities. Our approach is that of Chandra and Davis (1978), but we avoid a continuous-dependence assumption imposed in that paper and in Conway and Smoller (1977a).

For two vectors a, $b \in \mathbb{R}^n$, the symbol $a < b$ means $a_i < b_i$ for all i.

Let E_i, $i = 1, \ldots, n$, be uniformly elliptic linear second order partial differential operators in m space variables,

$$E_i z \equiv \sum_{j,k}^{m} a_{jki}(x,t)\partial_{x_j}\partial_{x_k} z + \sum_{j}^{m} b_{ji}(x,t)\partial_{x_j} z,$$

with uniformly bounded coefficients, defined in a closed space-time domain $\bar{Q}$, where $Q = \Omega \times (0,T)$ for some $T > 0$, and for some (possibly unbounded) domain $\Omega \subset \mathbb{R}^m$.

Let $F(x,t,u)$ and $\underline{F}(x,t,u)$, defined for $(x,t) \in \bar{Q}$, $u \in \mathbb{R}^n$, satisfy $F(x,t,u) \geq \underline{F}(x,t,u)$, with $\underline{F}$ Lipschitz continuous in u, uniformly for $(x,t) \in \bar{Q}$. For each j, let $\underline{F}_j$ be nondecreasing in each component u_k with $k \neq j$.

THEOREM 5.1. Let E_i, F, and $\underline{F}$ be as described above. Let u and $\underline{u}$ be continuous functions from $\bar{Q}$ into $\mathbb{R}^n$, C_2 in Q, bounded, and satisfying (for each $i = 1, \ldots, n$)

$$\partial_t u_i - E_i u_i \geq F_i(x,t,u),$$

$$\partial_t \underline{u}_i - E_i \underline{u}_i \leq \underline{F}_i(x,t,\underline{u}),$$

$$u \geq \underline{u} \quad \text{for} \quad t = 0,$$

$u \geq \underline{u}$ or $\dfrac{\partial u}{\partial \nu} \geq \dfrac{\partial \underline{u}}{\partial \nu}$ for $x \in \partial\Omega$, where $\dfrac{\partial}{\partial \nu}$ is an outward directional derivative.

Then $u \geq \underline{u}$ in $\bar{Q}$.

<u>Proof</u>: For a small parameter $\varepsilon > 0$, let

$$v_i = \underline{u}_i - \varepsilon(1 + 2Lt) \quad \text{for all} \quad i$$

(symbolically, we write $v = \underline{u} - \varepsilon(1 + 2Lt)$, where L is the Lipschitz constant for $\underline{F}$). Then for each i (suppressing dependence of $\underline{F}$ on (x,t)),

$$\partial_t v_i - E_i v_i = \partial_t \underline{u}_i - E_i \underline{u}_i - 2\varepsilon L$$

$$\leq \underline{F}_i(v + \varepsilon(1 + 2Lt)) - 2\varepsilon L$$

$$\leq \underline{F}_i(v) + L\varepsilon(1 + 2Lt) - 2\varepsilon L$$

$$\leq \underline{F}_i(v)$$

for $t \leq \frac{1}{2L}$.

Let $w = u - v$. For each j,

$$\partial_t w_j - E_j w_j = \partial_t u_j - E_j u_j - \partial_t v_j + E_j v_j$$

$$= F_j(x,t,u) - F_j(x,t,v)$$

$$\geq \underline{F}_j(x,t,u) - \underline{F}_j(x,t,v). \tag{5.1}$$

Let $\tau \in (0,\frac{1}{2L}]$ be such that $w \geq 0$ for $t \in [0,\tau]$. Such a τ exists, because $w \geq \varepsilon > 0$ for $t = 0$. Let $\hat{u}^j$ be the collection of the $n - 1$ components u_k with $k \neq j$, and write $F(u) = F(\hat{u}^j, u_j)$. Then by hypothesis, $\underline{F}_j$ is nondecreasing in $\hat{u}^j$, so

$$\underline{F}_j(x,t,u) = \underline{F}_j(x,t,\hat{u}^j,u_j) \geq \underline{F}_j(x,t,\hat{v}^j,u_j) \geq \underline{F}_j(x,t,v) - L(u_j - v_j),$$

for $t \in [0,\tau]$. From this and (5.1), we obtain that

$$\partial_t w_j - E_j w_j + L w_j \geq 0, \quad t \in [0,\tau].$$

By the maximum principle for scalar parabolic equations, we obtain that $w_j > 0$ for all j:

$$w > 0 \quad \text{for} \quad t \in [0,\tau].$$

In short, $\tau \in [0,\frac{1}{2L}]$ and $w \geq 0$ on $[0,\tau]$ imply $w > 0$ on $[0,\tau]$. It is therefore easy to see that $w \geq 0$ on $[0,\frac{1}{2L}]$. (The proof is by contradiction; or, let I be the set of values of t in $[0,\frac{1}{2L}]$ for which $w \geq 0$. The result just proved shows I is open relative to the basic interval $[0,\frac{1}{2L}]$, continuity of w shows it to be closed, and it is nonempty, so must be the entire interval.)

Letting $\varepsilon \to 0$, we obtain $u \geq \underline{u}$ for $t \in [0,\frac{1}{2L}]$. Continuing this process, we obtain the same inequality on $[\frac{1}{2L}, \frac{2}{2L}]$, and eventually on $[0,T]$. This completes the proof.

Of course, the analogous result holds with a supersolution $\bar{u}$ and corresponding $\bar{F}$.

This raises the question as to how to construct useful sub-and super-reaction functions $\underline{F}$ and/or $\bar{F}$, with sub- and super-solutions $\underline{u}$ and/or $\bar{u}$. Of course if the F_i are already nondecreasing in u_j, $j \neq i$, then we may take $\underline{F} = \bar{F} = F$. Otherwise, it has been productive to define

$$\underline{F}_j(x,t,u) \equiv \inf_{v \geq u} F_j(x,t,v_1,\ldots,v_{j-1},u_j,v_{j+1},\ldots,v_n), \tag{5.2a}$$

and

$$\bar{F}_j(x,t,u) \equiv \sup_{v \leq u} F_j(\ldots). \tag{5.2b}$$

Clearly, when u is increased, the infemum is taken over a smaller set, and the supremum over a larger set, so the values of $\underline{F}_j$ and $\bar{F}_j$ do not decrease. It is not hard to see that $\underline{F}$ and $\bar{F}$ are Lipschitz in u, provided F is.

If the intent is to bound the solution u simultaneously between a subsolution $\underline{u}$ and a supersolution $\bar{u}$, then the infemum above need be taken only over the set $u \leq v \leq \bar{u}$, and the supremum only over the set $\underline{u} \leq v \leq u$.

The following is a simple "invariant rectangle" result:

COROLLARY 5.2. Let $a < b$ be two constant vectors in $\mathbb{R}^n$ such that

$$F_j(x,t,u) \geq 0 \quad \text{for} \quad u_j = b_j, \quad a \leq u \leq b, \quad (x,t) \in \bar{Q},$$

and

$$F_j(x,t,u) \leq 0 \quad \text{for} \quad u_j = a_j, \quad a \leq u \leq b, \quad (x,t) \in \bar{Q}.$$

Let u be a solution of

$$\frac{\partial u_i}{\partial t} - E_i u_i = F_i, \quad i = 1, \ldots, n,$$

with $a \leq u \leq b$ for $t = 0$, and $a \leq u \leq b$ or $\frac{\partial u}{\partial \nu} = 0$ for $x \in \partial\Omega$.

Then $a \leq u \leq b$ for all (x,t) in Q.

In words, the rectangle $a \leq u \leq b$ is invariant for the reaction-diffusion system.

Proof: We set $\underline{u} = a$, $\bar{u} = b$, and define $\underline{F}$ and $\bar{F}$ as outlined above. The result follows from Theorem 5.1. Theorem 1.3 is an example of the corollary.

Note that containment in the rectangle constitutes an a priori bound which implies that a solution u with initial and boundary data in the rectangle exists globally. Much more general global existence theorems were obtained by Amann.

The following corollary is immediate.

COROLLARY 5.3 (Conway and Smoller 1977c). Let a, b, F, and u satisfy the hypotheses of Corollary 5.2 with u satisfying a homogeneous Neumann condition on $\partial\Omega$. Let F be independent of x, and let $\underline{F}$ and $\bar{F}$ be as defined in (5.2), with the substitutions $\inf_{u \leq v \leq b}$ and $\sup_{a \leq v \leq u}$. Let $\underline{U}(t)$ be defined by

$$\frac{d\underline{U}}{dt} = \underline{F}(t,\underline{U}), \quad \underline{U}(0) = \underline{U}_o$$

with an analogous definition for $\bar{U}(t)$. Here $a \leq \underline{U}_o \leq \bar{U}^o \leq b$.

If $\underline{U}_o \leq u(x,0) \leq \bar{U}^o$ for all $x \in \bar{\Omega}$, then $\underline{U}(t) \leq u(x,t) \leq \bar{U}(t)$ for all $(x,t) \in \bar{Q}$.

5.2 An example from ecology

As an example to show how the above Corollary 5.3 can establish stability of rest states, we look at the ecological equations governing the dynamics of two interacting species, with diffusion:

$$
\begin{aligned}
u_t &= D_1 \nabla^2 u + uM(u,v) \\
v_t &= D_2 \nabla^2 v + vN(u,v).
\end{aligned}
\tag{5.3}
$$

Our discussion is based on Conway and Smoller (1977a,b). Here u and v are defined for x in all space, or in a bounded domain Ω on the boundary of which Neumann boundary conditions are imposed.

In the present setting, Corollary 5.3 gives the following result. Consider a rectangle $\Sigma = \{a < U < b\}$ in the positive quadrant of the (u,v) plane (here $U = (u,v)$). According to Corollary 5.2, Σ will be invariant if $U(x,0) \in \bar{\Sigma}$,

$$M(a_1,v) \geq 0, \quad M(b_1 v) \leq 0,$$

$$N(u,a_2) \geq 0, \quad \text{and} \quad N(u,b_2) \leq 0 \quad \text{for} \quad U \in \bar{\Sigma}.$$

Let

$$M^+(u,v) = \sup\{M(u,\theta),\ a_2 \leq \theta \leq v\},$$

$$N^+(u,v) = \sup\{N(\theta,v),\ a_1 \leq \theta \leq u\},$$

$$M^-(u,v) = \inf\{M(u,\theta),\ v \leq \theta \leq b_2\},$$

$$N^-(u,v) = \inf\{N(\theta,v),\ u \leq \theta \leq b_1\}.$$

Let $\underline{u}(t)$, $\underline{v}(t)$, $\bar{u}(t)$, $\bar{v}(t)$ satisfy

$$\frac{d\bar{u}}{dt} = \bar{u}M^+(\bar{u},\bar{v}),$$

$$\frac{d\bar{v}}{dt} = \bar{v}N^{+}(\bar{u},\bar{v}),$$

with similar equations for $\underline{u}$ and $\underline{v}$. The result is that if $\underline{u}(0) \leq u(x,0) \leq \bar{u}(0)$, $\underline{v}(0) \leq v(x,0) \leq \bar{v}(0)$, $x \in \Omega$, where $(\underline{u}(0),\underline{v}(0)) \in \bar{\Sigma}$, $(\bar{u}(0),\bar{v}(0)) \in \bar{\Sigma}$, then

$$\underline{u}(t) \leq u(x,t) \leq \bar{u}(t); \quad \underline{v}(t) \leq v(x,t) \leq \bar{v}(t) \tag{5.4}$$

for all $t \geq 0$.

We assume that the corresponding kinetic equations have a stable equilibrium at $(p,0)$ for some $p > 0$. This equilibrium represents a state in which the second species has become extinct. The Jacobian matrix at $(p,0)$ is

$$J = \begin{pmatrix} pM_u(p,0) & pM_v(p,0) \\ 0 & N(p,0) \end{pmatrix},$$

whose eigenvalues are $\lambda_1 = pM_u(p,0)$, $\lambda_2 = N(p,0)$. Stability implies they are nonpositive, but we assume they are, in fact, strictly negative:

$$M_u(p,0) < 0 \tag{5.5}$$

$$N(p,0) < 0. \tag{5.6}$$

Our question is, does this imply that $(p,0)$ is also a stable state for the reaction-diffusion system (5.3)?

Consider the curve $M(u,v) = 0$. It contains $(p,0)$, since the latter is a rest state, and near that point, may be expressed in the form $v = v(u)$, if M is monotone in v. Let us assume, in fact, that

$$M_v(p,0) \neq 0.$$

We consider two cases:

(a) $M_v(p,0) > 0$. This implies that increasing v enhances the growth of u, so that v is a species beneficial to u. Differentiating $M(u,v(u)) = 0$, we find $M_u + M_v \frac{dv}{du} = 0$, so that $dv/du > 0$, and the curve $v(u)$ is as shown:

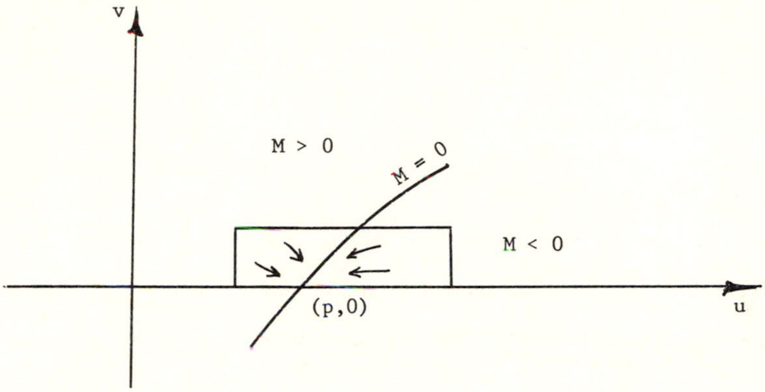

Let Σ be the small rectangle shown, resting on the u axis. The signs of M being as indicated, it can be seen that Σ is an invariant rectangle, if it contains the initial data.

Let us compute the vector fields $M^{\pm}$, $N^{\pm}$. Since M is increasing in v, we have $M^{+}(u,v) = M(u,v) = M^{-}(u,v)$, and of course $N^{\pm}(u,v) < 0$. Therefore in all cases, $M^{\pm}$ and $N^{\pm}$ have the same signs as M and N, respectively. It can be seen (see the arrows in the above figure) that the flows generated by (M^{+},N^{+}) and (M^{-},N^{-}) converge to the point $(p,0)$.

We take $(\underline{u}(0),\underline{v}(0))$ and $(\bar{u}(0),\bar{v}(0))$ to be the lower left and upper right corners of Σ. Then

$$\lim_{t\to\infty} \underline{u}(t) = \lim_{t\to\infty} \bar{u}(t) = p,$$

$$\underline{v}(t) \equiv \lim_{t\to\infty} \bar{v}(t) = 0.$$

By the inclusion relation (5.4), we have that $(u(x,t),v(x,t))$ converges uniformly to $(p,0)$ as $t \to \infty$, provided that $(u(x,0),v(x,0)) \in \Sigma$ for all x. This is the stability result we are seeking.

(b) $M_v(p,0) < 0$. This is the situation for predator-prey and competing species models: v inhibits the growth of u. In this case, we have $dv/du < 0$, as in the following picture:

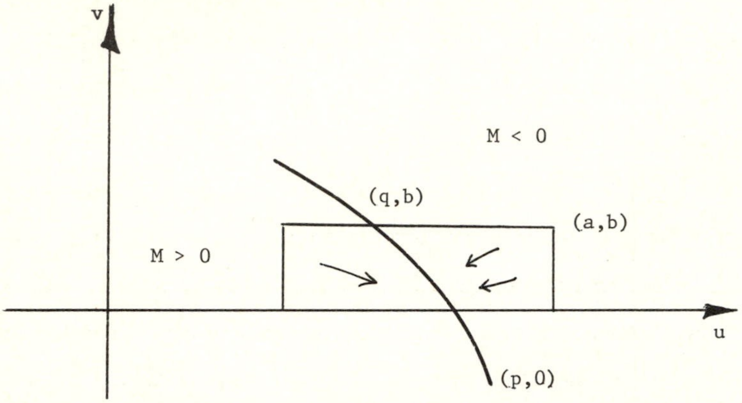

Since M is decreasing as a function of v in Σ, we find

$$M^+(u,v) = M(u,0); \quad M^-(u,v) = M(u,b).$$

As before, let $(\underline{u}(0),\underline{v}(0))$ and $(\bar{u}(0),\bar{v}(0))$ be the same two corners of Σ. Let (q,b) be the point of intersection of the curve $M = 0$ with the top of Σ (see the diagram). Then $M^-(u,v)$ vanishes for $u = q$, and so

$$\lim_{t\to\infty} \underline{u}(t) = q.$$

However, clearly $\bar{u}(t) \downarrow p$, $\underline{v}(t) \equiv 0$, $\bar{v}(t) \downarrow 0$. Let t_1 be such that $\bar{v}(t_1) = b/2$. Stop the process at that time, and construct another rectangle Σ_1 with opposite corners $(\underline{u}(t_1),0)$ and $(\bar{u}(t_1),b/2)$. With reference to Σ_1, we may set up another maximal and minimal field, and apply Corollary 5.3 again. In this case $\lim_{t\to\infty} \underline{u}(t) = q_1$, where $(q_1,b/2)$ is the intersection of the isocline $M = 0$ with the top of Σ_1. Of course, $q_1 > q$.

Continue the new process until such a time that $\bar{v}(t_2) = b/4$, then do as before. This way, we obtain a nested sequence of invariant rectangles converging to $(p,0)$, such that each one will eventually contain $(u(x,t),v(x,t))$ for all x. Therefore (u,v) converges uniformly to $(p,0)$ as before, so that $(p,0)$ is asymptotically stable for the reaction-diffusion system.

In the case of Volterra-Lotka equations with diffusion in a bounded domain, possibly with many species, various authors have proved convergence of general solutions to uniform stable stationary states (when one exists). Lyapunov functional methods may be used to advantage here. See Rothe (1976), A. Hastings (1977), Williams and Chow (1978), and de Mottoni and Rothe (1978), for instance.

For any x-independent RD system, and a solution satisfying zero Neumann data on the boundary of a bounded domain, criteria were given by Conway, Hoff, and Smoller (1978) and Othmer (1977) for the solution to approach a function of t alone as $t \to \infty$. Of course, the limit function will be a solution of the associated kinetic equations (1.25). The criterion given by Conway et al. requires the solution to be in an invariant rectangle (as in Corollary 5.2), and a certain parameter to be small enough. This last will be the case for a sufficiently small domain or sufficiently large diffusion coefficients.

Klaasen and Troy (in preparation) have used comparison techniques to obtain asymptotic speed of propagation results for a certain wave front whose existence was established by Troy (1978b).

6. SYSTEMS: LINEAR STABILITY TECHNIQUES

6.1 Stability considerations for nonconstant stationary solutions and traveling waves

The application given in the preceding Section 5.2 essentially is a case in which adding diffusion to a system of kinetic equations does not destroy or change the character of a stable stationary state. As mentioned, similar results (Conway, Hoff, and Smoller 1978) state that if the spatial domain Ω of the population is small enough, or if the diffusion coefficients are large enough, then the long-time behavior of solutions of the reaction-diffusion system

$$\frac{\partial u}{\partial t} = D\Delta u + F(u) \tag{6.1}$$

under homogeneous Neumann boundary conditions is essentially the same as corresponding solutions of the associated kinetic system. In this chapter, on the other hand, we begin to explore the possibility of stable structures which do depend on the mechanism of diffusion for their existence. We give some general remarks about linear stability, and in Section 6.2 and later chapters show conditions under which such stable "dissipative" structures may be constructed. First, it should be mentioned that Bardos and Smoller (1978) have delineated a class of RD systems for which stationary stable dissipative structures do not exist.

Let $u = \phi(x)$ be a stationary solution of (6.1). If the domain Ω of the solution is all space, then more generally we may allow $u = \phi(\nu\cdot x - ct)$ to be a plane wave; $\phi(z)$ will then be a stationary solution of the modified equation

$$u_t - cu_z + Du_{zz} - F(u) = 0,$$

and the stability analysis we describe applies to this modified system as well. But for simplicity, we restrict attention to stationary solutions of (6.1).

Let S denote the differential operator obtained by formally linearizing the right hand side of (6.1) about the given solution ϕ:

$$S(\bar{u}) \equiv D\nabla^2\bar{u} + \frac{\partial F}{\partial u}(\phi(x))\bar{u},$$

$\partial F/\partial u$ denoting the Jacobian matrix. We now interpret S as acting on a specific space of functions of x for $x \in \Omega$. If Ω has a boundary, the no-flux condition

$$\frac{\partial u}{\partial \nu} = 0, \quad x \in \partial\Omega \tag{6.2}$$

is imposed and serves to restrict the domain of definition of S. In some theories, S and the boundary condition are interpreted in generalized senses, but in the present exposition we interpret all differential operators in their classical sense. For definiteness, we consider S as an operator on the space $C^0(\bar{\Omega})$ of bounded vector functions, condinuous on $\bar{\Omega}$, with domain consisting of functions in $C^2(\bar{\Omega})$ satisfying the given boundary conditions.

We examine the spectrum $\Sigma(S)$. Recall it is the complement, in the complex plane, of the resolvent set, and the resolvent set is the set of complex numbers λ for which $S - \lambda I$ has a bounded inverse defined on the whole space.

Definition: ϕ is stable according to the linearized criterion (ℓ.c.) if $\Sigma(S)$ is in the negative $\frac{1}{2}$-plane, and is bounded away from the imaginary axis. ϕ is marginally stable (ℓ.c.) if $\Sigma(S)$ is in the negative closed $\frac{1}{2}$-plane, and is not bounded away from the imaginary axis. ϕ is unstable (ℓ.c.) if it contains a point in the right open $\frac{1}{2}$-plane.

These definitions would be considered justified if it could be shown that stability or instability (ℓ.c.) implies stability or instability in the more usual sense. This has been done, to a great extent. For example, Kielhöfer (1976) has, under wide conditions, proved that stability (ℓ.c.) implies asymptotic stability: a perturbation, small in the norm of the underlying space, generates a solution which approaches the given stationary solution as $t \to \infty$. He also showed that instability (ℓ.c.) implies instability in the usual sense.

Sattinger (1975, 1976, 1977) treated a case of marginal stability for problems in which the spatial domain is the whole real line. His underlying spaces are weighted L_∞ spaces. His crucial assumption is that $\Sigma(S)$ contains a simple eigenvalue at the origin, plus a part which is in the stable half-plane and bounded away from the imaginary axis. He makes an additional assumption about $\Sigma(S)$ which is probably true for most if not all reaction-diffusion equations of the type we are considering. Sattinger's conclusion is that a small perturbation of the given stationary solution $\phi(x)$ generates a solution which approaches $\phi(x - \delta)$ as $t \to \infty$ for some small shift δ.

Verifying the relevant properties of $\Sigma(S)$ is not an easy task, and for reaction-diffusion systems with $n > 1$, has only been achieved in a couple of very special cases (Feroe 1977, Fife 1977b).

Rigorous stability criteria in the same spirit as Sattinger's, but directed to general nerve signal equations, were developed by Evans (1972a,b, 1975).

6.2 Pattern stability for a class of model systems

Before giving an example of positive stability results for a class of systems, let us recall the negative picture for scalar equations. For problems on the whole real line, we have shown in Theorem 4.9 that there are no stable bounded stationary solutions except some of the constant solutions and the nonconstant monotone solutions (if they exist - but this would only be accidental). The same result is true for problems in one space variable on a bounded interval, no-flux boundary conditions being imposed at the endpoints. In that case, a variation of the same argument can be applied to show that no nonconstant stationary solutions can be stable. Alternatively, this can be shown (in the bounded interval case) by a linearized analysis, according to the criteria in Section 6.1.

Contrary to the scalar case, stable patterns can certainly exist for systems. The appearance of small amplitude solutions bifurcating from spatially constant solutions is well known and will be touched on in the next chapter. Now we show how to construct a class of model systems exhibiting stable large amplitude stationary solutions of a fairly general nature (Fife 1977b). The model systems will have two equations $(n = 2)$. C^0-stability or marginal stability can be established rigorously. Other methods (involving singular perturbations) can also be used to construct large amplitude patterns, for which a heuristic argument asserting stability can be given (see Chapter 8).

First, we consider problems on a finite interval, say $x \in [0,1]$, with no-flux boundary conditions $\underline{u}_x(0) = \underline{u}_x(1) = 0$, where $\underline{u} = (u,v)$. Suppose the diffusion coefficient of u to be 1, and let k be that of v. Thus, the type of system under consideration is

$$u_t = u_{xx} + f(u,v), \tag{6.3a}$$

$$v_t = kv_{xx} + g(u,v). \tag{6.3b}$$

Let $\phi(x)$ be a C^2 function on $[0,1]$ such that ϕ'' is a smooth function of ϕ (say $\phi'' = G(\phi)$), and $\phi'(0) = \phi'(1) = 0$. The former requirement is equivalent to saying that ϕ is even with respect to every relative maximum and minimum; i.e., if ϕ attains a maximum or minimum at x_1, then $\phi(x_1 + s) = \phi(x_1 - s)$ for every s such that $x_1 \pm s \in [0,1]$. An example is shown in the figure below.

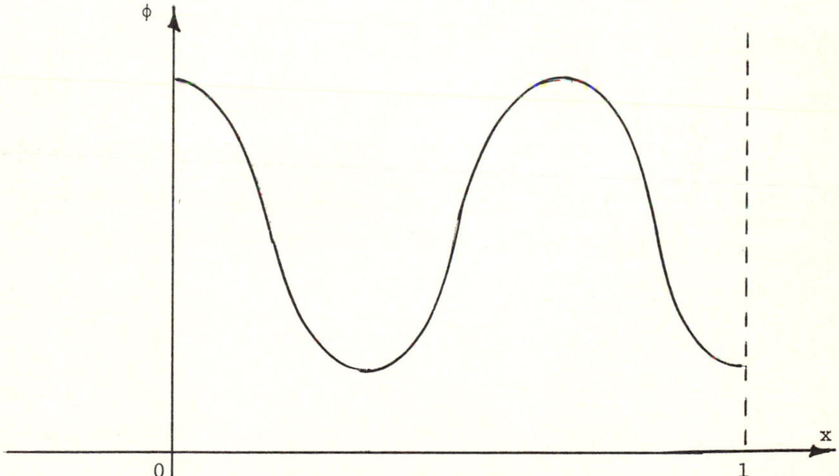

THEOREM 6.1. Let ϕ be as described, and let $\psi(x) = a\phi(x) + b$ for some constants a and b with $a > 0$. Then there are systems of the form (6.3) with $k > 1$ for which $u = \phi$, $v = \psi$ is a stable or, in exceptional cases, marginally stable solution.

For simplicity, we first consider the case $a = 1$, $b = 0$, so that $\psi \equiv \phi$.

We first note that if we were just looking for a system for which (ϕ,ϕ) is a solution (with no consideration of stability), the answer would be trivial:

$$u_t = u_{xx} - G(u),$$

$$v_t = v_{xx} - G(v).$$

But this system is uncoupled, and since ϕ is unstable as a solution of either single equation, (ϕ,ϕ) will be unstable for the system. So alter the system in such a way that (ϕ,ϕ) is still a solution. One way is the following:

$$u_t = u_{xx} - G(u) + \sigma(u - v) \tag{6.4a}$$

$$v_t = k[v_{xx} - G(v) + \sigma(u - v)], \tag{6.4b}$$

for some constants σ and k. We now see whether they can be adjusted so that (ϕ,ϕ) is not only a solution, but is in fact a stable one.

Let S be the linearization of the right hand side of (6.4) about $u = v = \phi$:

$$S\begin{pmatrix}\hat{u}\\ \hat{v}\end{pmatrix} = \begin{pmatrix}\hat{u}_{xx} - G'(\phi(x))\hat{u} + \sigma(\hat{u} - \hat{v})\\ \\ k[\hat{v}_{xx} - G'(\phi(x))\hat{v} + \sigma(\hat{u} - \hat{v})]\end{pmatrix} = \begin{pmatrix}L\hat{u} + \sigma(\hat{u} - \hat{v})\\ \\ k[L\hat{v} + \sigma(\hat{u} - \hat{v})]\end{pmatrix},$$

where $Lu \equiv u_{xx} - G'(\phi(x))u$. Interpret the operator L as acting on $C^o[0,1]$ with domain the class of C^2 functions satisfying no-flux boundary conditions. Interpret S as acting on a similar space of vector functions.

We need to investigate the spectrum $\Sigma(S)$. For this, we define

$$\varepsilon = 1/k,$$

$$P(\lambda,\mu) = \varepsilon\lambda^2 + \lambda[\sigma(1 - \varepsilon) - (1 + \varepsilon)\mu] + \mu^2,$$

$$\Lambda = \{\lambda : P(\lambda,\mu) = 0 \quad \text{for some} \quad \mu \in \Sigma(L)\}.$$

Our first result is that

$$\Sigma(S) \subset \Lambda. \tag{6.5}$$

To prove this, suppose that $\lambda \notin \Lambda$. Then $P(\lambda,\mu) \neq 0$ for all $\mu \in \Sigma(L)$, so $0 \notin P(\lambda,\Sigma(L))$. Now consider the fourth order differential operator $P(\lambda,L)$. It is known that for polynomials P, $\Sigma(P(L)) = P(\Sigma(L))$, so that in our case, we deduce $0 \notin \Sigma(P(\lambda,L))$. Therefore $P(\lambda,L)^{-1}$ exists as a bounded operator on $C^{o}[0,1]$. Given f and g in the latter space, we can solve the equations

$$(S - \lambda I)\begin{pmatrix} \hat{u} \\ \hat{v} \end{pmatrix} = \begin{pmatrix} f \\ g \end{pmatrix}$$

explicitly as follows:

$$\hat{u} = (L - \sigma - \varepsilon\lambda)P(\lambda,L)^{-1}f + \varepsilon\sigma P(\lambda,L)^{-1}g,$$

$$\hat{v} = -\sigma P(\lambda,L)^{-1}f + \varepsilon(L + \sigma - \lambda)P(\lambda,L)^{-1}g.$$

(This formula is like Cramer's rule for linear algebraic equations.) This shows that $\lambda \notin \Sigma(S)$, and this conclusion in turn proves (6.5).

Our problem has been reduced to determining Λ; for this, we need to know $\Sigma(L)$. Since L is a formally self-adjoint second order operator on a bounded interval, $\Sigma(L)$ will consist of an infinite but discrete set of simple real eigenvalues, bounded from above:

$$\Sigma(L) < M.$$

(Furthermore it contains a point $\mu > 0$, since we know $u = \phi$ is unstable as a solution of $u_t = u_{xx} - G(u)$.)

Let $\mu \in \Sigma(L)$. Suppose $\mu \neq 0$. Then the two roots λ_i of $P(\lambda,\mu) = 0$ have sum $-(1 - \varepsilon)\sigma + (1 + \varepsilon)\mu < -(1 - \varepsilon)\sigma + (1 + \varepsilon)M$ and product μ^2. The sum will therefore be negative if $0 < \varepsilon < 1$ and σ is large enough, and the product positive. Therefore $\text{Re}\ \lambda_i < 0$.

On the other hand if $\mu = 0$, then one of the roots is negative and the other zero.

We therefore obtain that Λ consists of discrete values in the negative half-plane, except possibly for a simple point at the origin (in case L has 0 as eigenvalue). The same will be true of $\Sigma(S)$, and so (ϕ,ϕ) will have the stability properties described in the theorem.

If $\psi = a\phi + b$, then to find the appropriate system (6.3), we first go through the above procedure to find a system which works when $\psi = \phi$, only use the symbol w in place of v. Then we effect the change of variable $w = av + b$.

The only requirements we have placed on the parameters of the problem were $k > 1$ (in (6.4b)) and σ sufficiently large. This means the inverse problem has a great many solutions.

Lefever, Herschkowitz-Kaufman, and Turner (1977) produced a RD system allowing analytical treatment and supporting stable dissipative structures. F. Rothe (1978) deduced the existence of such solutions for another model system.

Stable patterns for still other systems of reaction-diffusion equations had previously appeared in many computer simulations by various people at the University of Brussels (see references in Nicolis and Prigogine (1976)), and by Gierer and Meinhardt (1972, 1974); see also Gierer (1976) and Meinhardt (1974, 1976). Keener (1978) gave an asymptotic treatment of some of Gierer and Meinhardt equations. It is interesting that the model systems devised by Gierer and Meinhardt were of "activator-inhibitor" type, with the inhibitor diffusing more rapidly than the activator, and that in one sense that is also true of our system. We may define:

$$u \text{ is an activator if } f_u > 0 \text{ and } g_u > 0;$$

$$v \text{ is an inhibitor if } f_v < 0 \text{ and } g_v < 0.$$

For large enough σ, the model equations we have come up with are such that u and v are indeed an activator and inhibitor, respectively, and v's diffusivity is greater than that of u (since $k > 1$).

Much the same process can be repeated for problems on the infinite interval. For example, consider the "single peak" function

$$\phi(x) = \frac{6e^x}{(1 + e^x)^2} ,$$

whose graph looks like this:

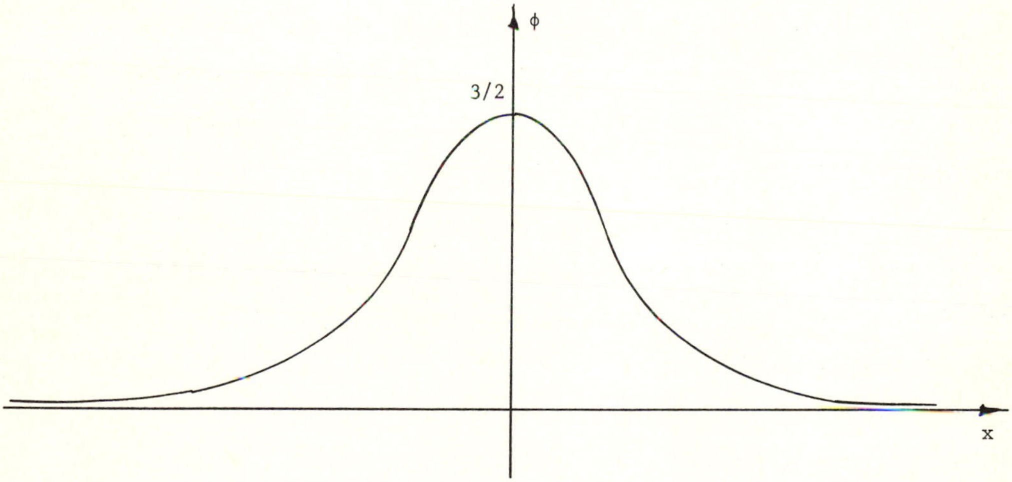

It satisfies $\phi'' = \phi - \phi^2$. But in general, given any bounded single-peak function satisfying $\phi'' = G(\phi)$ for some G with $G'(0) > 0$ and $\phi(\pm\infty) = 0$, the system (6.4) will support $u = \phi$, $v = \phi$ as a stable solution.

The argument to show this is very little changed from the foregoing. Now $\Sigma(L)$ is no longer discrete, but the continuous part is bounded away from 0. Since $\Sigma(L)$ always has 0 as an eigenvalue, $\Sigma(S)$ will have exactly the properties assumed by Sattinger (see Section 6.1). His theorem will therefore yield the stability of the given pattern.

Periodic patterns on the whole line may also be analyzed; but since the continuous part of the spectrum extends all the way to zero, Sattinger's theorem may no longer be applied; all we obtain is marginal stability according to the linearized criterion.

7. SYSTEMS: BIFURCATION TECHNIQUES

7.1 Small amplitude stationary solutions

In many reaction-diffusion problems, it is the case that a uniform solution exists, but is stable only for certain ranges of the parameters present in the equation. Then it is typically also the case that when the parameters assume values near the "transition" zone between stability and instability of the uniform solution, other nonuniform, but small amplitude, solutions exist as well. Sometimes they are stable, and thus represent new solutions to which stability has been transferred from the uniform solution. This appearance of new solutions is considered a bifurcation phenomenon, because in parameter-amplitude diagrams, new solutions branch off the known uniform solution at critical values of the parameters. The relevant mathematical background in bifurcation theory is available in a number of sources, including Sattinger (1971, 1973) and Crandall and Rabinowitz (1971, 1973, 1977).

Of course other types of "branching" of solutions also occur, besides those from uniform solutions, but we concentrate attention on the situation described. Moreover, we imagine only a single real parameter to be present. In this section we examine only the appearance of new stationary solutions.

Consider a system of the form (6.1) with one space variable, in which F depends also on a real parameter λ: $F = F(u,\lambda)$. Assume the system has a constant solution u_o for all values of λ: $F(u_o,\lambda) = 0$. By subtracting off u_o, we may suppose, for simplicity, that $u_o = 0$. Thus

$$F(0,\lambda) = 0.$$

We also assume, for simplicity, that the dependence on λ is only in the linear part of F. We may then write (6.1) in the form

$$y_t = Du_{xx} + (A + \lambda B)u + g(u), \tag{7.1}$$

with $g = O(|u|^2)$ as $|u| \to 0$ and A, B (as well as D) are matrices. We consider (7.1) for x on the bounded interval $[0,\pi]$ and apply no-flux boundary conditions $u_x(0,t) = u_x(\pi,t) = 0$.

We first define what we mean by loss of stability of the zero solution. Consider the corresponding linear problem

$$v_t = Dv_{xx} + (A + \lambda B)v, \quad x \in [0,\pi]; \quad v_x(0,t) = v_x(\pi,t) = 0. \tag{7.2}$$

For any integer n, it has solutions of the form

$$v = \Phi \cos nxe^{\sigma t};$$

in fact substituting this expression into (7.2), we have

$$(-n^2 D + (A + \lambda B) - \sigma I)\Phi = 0,$$

so that σ is an eigenvalue, with corresponding eigenvector Φ, of the matrix $H(n^2,\lambda)$ where we define

$$H(p,\lambda) \equiv -pD + A + \lambda B.$$

Moreover, the set of solutions of that form are a basis for the set of solutions of (7.2).

The linear criterion of stability of the zero solution is that all eigenvalues σ have negative real part, for all integers n. If there is an n such that $H(n^2,\lambda)$ has an eigenvalue with positive real part, the zero solution is unstable. Accordingly, we make the following hypothesis.

(1) $H(p,\lambda)$ has a unique eigenvalue with largest real part; call it $\sigma(p,\lambda)$; it is real and algebraically simple.

(2) There is a number λ_o and a $\delta > 0$ such that

(a) for $\lambda_o - \delta < \lambda < \lambda_o$, $\sigma(n^2,\lambda) < 0$ for all n;

(b) for $\lambda = \lambda_o$, there is an n_o such that $\sigma(n_o^2,\lambda_o) = 0$;

(c) for $\lambda_o < \lambda < \lambda_o + \delta$, $\sigma(n_o^2,\lambda) > 0$, whereas $\sigma(n^2,\lambda) < 0$ for $n \neq n_o$.

(3) $\frac{\partial}{\partial\lambda}\sigma(n_o^2,\lambda)\Big|_{\lambda=\lambda_o} > 0.$

Under these hypotheses, we say that instability sets in at the critical value λ_o, and that the basic unstable mode is $\Phi \cos n_o x$. If $n_o = 0$, the kinetic equations ((7.1) with $D = 0$) become unstable at the same time as (7.1); then diffusion has no influence on stability. On the other hand, if $n_o \neq 0$, then the zero solution becomes unstable to some nonuniform perturbations, but remains stable to x-independent perturbations. For a discussion of this seemingly paradoxical fact that diffusion may induce instability, see Segel and Jackson (1972) and Segel and Levin (1975). Incidentally, for diagonal matrices D, it is impossible for n_o to be different from zero when all diffusion coefficients are equal.

Example: Let

$$D = \begin{pmatrix} 1 & 0 \\ 0 & 2 \end{pmatrix}, \quad A = \begin{pmatrix} 0 & -4 \\ 8 & -6 \end{pmatrix}, \quad B = \begin{pmatrix} 1 & 0 \\ 0 & 0 \end{pmatrix}.$$

Then

$$H(p,\lambda) - \sigma I = \begin{pmatrix} -p + \lambda - \sigma & -4 \\ 8 & -2p - 6 - \sigma \end{pmatrix}.$$

Setting the determinant equal to zero and solving for the largest root σ, we have

$$\sigma = \frac{1}{2}(\lambda - 6 - 3p) + \frac{1}{2}\sqrt{(\lambda - 6 - 3p)^2 - 4Q(p,\lambda)},$$

where $Q(p,\lambda) = (p - \lambda)(2p + 6) + 32$. Clearly

$$\text{Re}\,\sigma < 0 \quad \text{if and only if} \quad (1)\ \lambda - 6 - 3p < 0$$

$$\text{and} \quad (2)\ Q(p,\lambda) > 0.$$

Let λ range over positive values. As λ increases, the first inequality is first violated for $p = 0$, $\lambda = 6$. The second is violated when $Q_m(\lambda) = \text{Min}_{p=n^2} Q(p,\lambda) = 0$. But $Q_m(\lambda) = -\frac{1}{2}(\lambda + 3)^2 + 32$, and is attained for $p = \frac{1}{2}\lambda - \frac{3}{2}$. So the second inequality is first violated for $\lambda = 5$, $p = 1 = n_o^2$. This occurs before the first is violated, so in our example,

$$\lambda_o = 5, \quad n_o = 1.$$

For simplicity, in the following we take $\lambda_o = 0$. Let Φ be the nullvector of $H(n_o^2,0)$:

$$H(n_o^2,0)\Phi = 0,$$

and Ψ that of the transpose:

$$H^*(n_o^2,0)\Psi = 0.$$

By assumption, Φ and Ψ are unique, except for a constant factor. The functions

$$\phi = \Phi \cos n_o x, \quad \psi = \Psi \cos n_o x$$

will be the null functions of the operator L and its adjoint, respectively, where

$$Lv \equiv Dv_{xx} + Av, \quad L^*v = D^*v_{xx} + A^*v.$$

We set $(u_1,u_2) \equiv \int_0^\pi u_1(x)\cdot u_2(x)dx$, and normalize to $\|\phi\|^2 = (\phi,\phi) = \|\psi\|^2 = 1$.

We are posing the following problem: Find nontrivial solutions $u(x)$ of

$$Du_{xx} + (A + \lambda B)u + g(u) = 0; \quad \text{bdry conds.} \tag{7.3}$$

Any solution can be written as a multiple of ϕ plus a function orthogonal to ϕ in the sense of the above scalar product. We let ε be the scalar multiple of ϕ in this, and write the eventual solution u in the form

$$u = \varepsilon(\phi + w), \quad (w,\phi) = 0. \tag{7.4}$$

We shall seek solutions of (7.3) in this form, where w is a function of ε (as well as of x) for small ε, and $\lambda = \lambda(\varepsilon)$. Putting (7.4) into (7.3) and dividing by ε, we obtain

$$\begin{gathered} Lw + \lambda B(\phi + w) + \varepsilon h(\phi + w,\varepsilon) = 0. \\ w_x(0) = w_x(\pi) = 0, \quad (w,\phi) = 0, \end{gathered} \tag{7.5}$$

where $h(v,\varepsilon) \equiv \varepsilon^{-2}g(\varepsilon v)$. Here h is regular in ε at $\varepsilon = 0$, because g is quadratic or higher order in u.

We show how a solution of (7.5) in the form of a formal power series expansion may be obtained.

First, we remark that the equation

$$Lu = f; \quad \text{bdry conds.}$$

may be solved for u if and only if $(f,\psi) = 0$. This is because L is a Fredholm operator with nullspace spanned by ϕ, and ψ is the nullvector of the adjoint operator L^*.

Finally, if this orthogonality condition is fulfilled, then there is a unique solution satisfying $(u,\phi) = 0$. This is because the totality of solutions is the set $\{u_o + C\phi\}$, where u_o is any given solution and C is an arbitrary constant. Requiring $(u,\phi) = 0$ specifies C uniquely as $C = -(u_o,\phi)$.

Applying this orthogonality condition to (7.5), we obtain

$$\lambda(B(\phi + w),\psi) + \varepsilon(h(\phi + w,\varepsilon),\psi) = 0. \tag{7.6}$$

This latter equation indicates that λ will be of order ε or smaller as $\varepsilon \to 0$, and (7.5) in turn indicates the same is true of w.

To obtain our first approximation, then, we discard all higher order terms in (7.5) and (7.6), to obtain

$$Lw + \lambda B\phi + h(\phi,0) \cong 0,$$

$$\lambda(B\phi,\psi) + \varepsilon(h(\phi,0),\psi) \cong 0,$$

hence, provided

$$b \equiv (B\phi,\psi) \neq 0, \tag{7.7}$$

we have

$$\lambda = -\frac{\varepsilon}{b}(h(\phi,0),\psi),$$

$$w = -\varepsilon L^{-1}[h(\phi,0) - \frac{1}{b}(h(\phi,0),\psi)B\phi].$$

It turns out, however, that we need not assume (7.7) because it follows already from our original assumptions. To see this, we note that assumption (1) implies σ and Φ are smooth functions of p and λ. We differentiate the equation

$$(H(p,\lambda) - \sigma(p,\lambda)I)\Phi(p,\lambda) = 0$$

with respect to λ to obtain

$$(H - \sigma I)\Phi_\lambda = -(H_\lambda - \sigma_\lambda I)\Phi. \tag{7.8}$$

Now set $p = n_o^2$, $\lambda = \sigma = 0$, and recall that 0 is then an algebraically simple eigenvalue of H. This implies the equation has a solution Φ_λ if and only if the right hand side of (7.8),

$$(H_\lambda - \sigma_\lambda I)\Phi = B\Phi - \sigma_\lambda \Phi$$

is orthogonal to Ψ:

$$B\Phi \cdot \Psi = \sigma_\lambda \Phi \cdot \Psi. \tag{7.9}$$

Since we know the equation certainly does have a solution Φ_λ, it follows that (7.9) holds. And since the algebraic simplicity implies $\Phi \cdot \Psi \neq 0$, and condition (3) says $\sigma_\lambda \neq 0$, we conclude that $B\Phi \cdot \Psi \neq 0$. But this is equivalent to (7.7).

The above first approximation shows λ and w to be first order in ε, for small ε. Actually, however, λ is always second order, because (as we shall see),

$$(h(\phi,0),\psi) = 0. \tag{7.10}$$

The second and all higher order approximations to w and λ may be obtained in a straightforward way; for example, to obtain the second order terms, we set $w \sim \varepsilon w_1 + \varepsilon^2 w_2$, $\lambda \sim \varepsilon\lambda_1 + \varepsilon^2\lambda_2$, put these into (7.5), (7.6), discard terms of order in ε higher than two, and solve for λ_2 and w_2 (as we remarked above, $\lambda_1 = 0$).

Finally, we shall show that (7.10) holds. For this we observe that $h(\phi,0)$ is quadratic in ϕ since g was, and so

$$h(\phi,0) = h^* \cos^2 n_o x$$

for some vector h^*. Therefore

$$(h(\phi,0),\psi) = h^*\cdot\Psi \int_0^\pi \cos^3 n_o x dx = 0.$$

The expansion for λ is therefore of the form

$$\lambda = \varepsilon^2\lambda_2 + \varepsilon^3\lambda_3 + \dots .$$

The relation between ε and λ can be shown as follows:

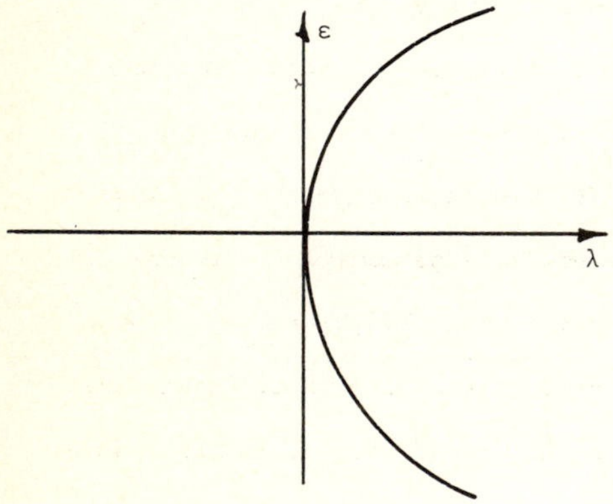

$(\lambda_2 > 0)$

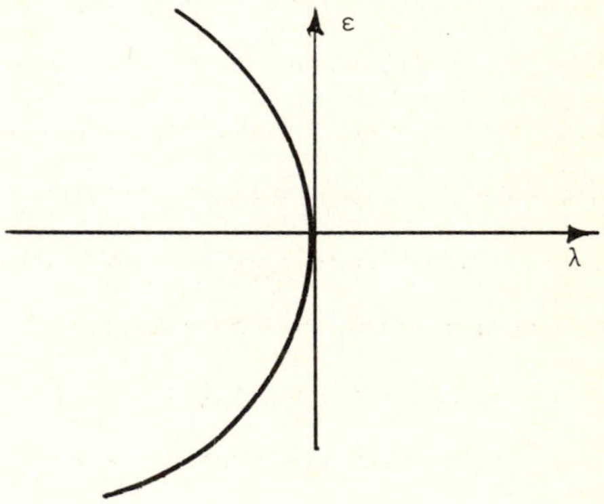

$(\lambda_2 < 0)$

The curving branch also represents, of course, the branch of nontrivial solutions of (7.3); we merely note that ε is a measure of the magnitude of u.

A standard result in bifurcation theory (see, for example, Crandall and Rabinowitz (1973)) tells us that solutions on a curve bifurcating to the right are stable, whereas those bifurcating to the left are unstable. So the small nontrivial solutions we have constructed are stable if $\lambda_2 > 0$, and unstable if $\lambda_2 < 0$.

The above outlines a procedure for obtaining approximate solutions of (7.3) in the form of power series in ε. As far as the existence of exact solutions, this result can be achieved by an application of the implicit function theorem to the pair of equations (7.5), (7.6), to yield w and λ as functions of ε.

The nature of the nonuniform solution which arises through bifurcation is influenced, at least in part, by the boundaries at $x = 0$ and $x = \pi$. Specifically, the principle unstable mode is such that it must satisfy the prescribed boundary conditions; this limits the wave number n_o of the mode to be an integer, and this further influences the critical value of λ_o at which the bifurcation takes place. It is of interest to study problems without boundary, with a view to the possible appearance of small amplitude patterns, again through bifurcation, which are due solely to the internal processes of reaction and diffusion.

This can be done (Fife 1977a). Assumptions much like (1)-(3) on pages 144-145 are made; the difference is that in (2) and (3), n^2 and n_o^2 are replaced by p and p_o, which are allowed to vary continuously, rather than be squares of integers. Assumption (2c) is replaced by the condition that for $\lambda_o < \lambda < \lambda_o + \delta$, the set of values of p for which $\sigma(p,\lambda) > 0$ is a finite interval. Periodic solutions are sought with arbitrary wave number $k = \sqrt{p}$. They are found to exist, not only for all small ε, but for all values of p near enough to p_o. Thus, a two-parameter family of periodic patterns is found. The parameter λ is also a function of these two parameters: $\lambda = \lambda(\varepsilon, p - p_o)$, with $\lambda(0,0) = \lambda_o$.

The bifurcation analysis in the infinite habitat case is more difficult than that in the previous case, because L now has a two-dimensional nullspace. Nevertheless, an argument involving the invariance of the differential equation when x is replaced by $-x$ can be used to show that every periodic solution must be a translate of an even function. And when the bifurcation analysis is restricted to spaces of even functions, the degeneracy associated with the two-dimensional nullspace disappears.

The question of stability of bifurcating structures on the entire line is, again contrary to the finite domain case, not an easy matter. Multiple scale techniques, as in Newell and Whitehead (1969), Fife (1978a), Ermentrout (1979), offer the most promise at this point. In recent papers, Cohen, Hagan, and Simpson (1978a, b) use multiple scaling for systems described below. They make a statement to the effect that bifurcating periodic stationary solutions (and wave trains) are unstable. But their analysis does not apply to the solutions constructed in Fife (1977a), some of which are no doubt stable.

7.2 Small amplitude wave trains

The search for wave train solutions of RD systems was spurred by their exhibition in the laboratory by Zaikin and Zhabotinskiĭ (1970) (see also Zhabotinskiĭ and Zaikin (1973)). A good exposition of theory and observation here is by Tyson (1976).

When one is looking for wave trains, it is inappropriate to work in a bounded domain, so we suppose the solution u of (7.1) to be defined for all x and t. Hypotheses analogous to (1)-(3) lead to the appearance of small amplitude periodic traveling waves. The difference is the following: In the last section we supposed that the eigenvalue with maximal real part is real and crosses the imaginary axis as λ increases through zero; but now we suppose it to be complex, and again to cross the imaginary axis with its complex conjugate as λ increases through zero.

Let $\sigma(p,\lambda)$ now denote the real part of this eigenvalue, which is again assumed to be algebraically simple. Then we continue to assume (2) and (3) on pages 144-145, with n^2 and n_o^2 replaced by p and p_o, respectively. Under these hypotheses there exists for λ near λ_o, a bifurcating two-parameter family of wave trains $u = U(kx + \omega t)$ with k near $\sqrt{p_o}$ and ω near ω_o = imaginary part of the eigenvalue of $H(p_o,\lambda_o)$. See Fife (1978b) for some more details.

Different small amplitude wave trains, which are clearly unstable, were constructed by Kopell and Howard (1973) and Ortoleva and Ross (1974). Cohen, Hagan, and Simpson (1978a,b) construct analogous unstable bifurcating wave trains for more general equations and systems allowing dependence on past history.

7.3 Bibliographical discussion

Turing's celebrated paper (1953) involved essentially a bifurcation type of analysis, because he sought the appearance of patterned structures through the onset of instability of uniform solutions of a RD system. He also surmised the appearance of waves (traveling in a ring) this way. His motivation was interest in biological morphogenesis, as has been the case for many later researches into pattern formation phenomena. (It is increasingly clear that many mechanisms play a role in morphogenesis; the role of pattern formation by RD systems is not yet so clear.)

Turing's ideas have been carried forward by a number of researchers since the late 1960's; some of the principal references are Gmitro and Scriven (1966), Prigogine and Nicolis (1967), Prigogine and Lefever (1968), Othmer and Scriven (1969), Glansdorff and Prigogine (1971), Ortoleva and Ross (1974), Boa (1974), Othmer (1975), Auchmuty and Nicolis (1975, 1976, 1977), Herschkowitz-Kaufman (1975), Balshev and Degn (1975), Boa and Cohen (1976), Nicolis and Prigogine 1977), Othmer (1977), and many others.

Levin (1978) has given a discussion of the role of bifurcation phenomena in the analysis of ecological communities; in this connection, see also Levin (1976), Segel and Jackson (1972), and Segel and Levin (1976).

A variety of other interesting branching phenomena for RD systems were explored by Keener (1976); see also Othmer (1977). De Simone, Beil, and Scriven (1973) discovered bifurcating target patterns and spirals in a membrane model. See also Ortoleva and Ross (1974). Ibañez and Velarde (1978) explored a phenomenon of bifurcation into patterned state on a sphere. Larson (1977) found bifurcating pulse solutions (unstable), and Fife (1978b) showed how to construct bifurcating wave fronts.

Linear stability analyses, closely related to bifurcation, were pursued by Casten and Holland (1977), and by de Mottoni and Tesei (1979) for a model arising in nuclear reactor theory. Chafee and Infante (1974) gave a thorough treatment of the scalar case on a finite interval.

"Hopf bifurcation" refers to the appearance, when some parameter crosses a critical value, of small time-periodic solutions. It is a common phenomenon for ordinary differential equations. Analogous phenomena for reaction-diffusion systems have been investigated by Auchmuty and Nicolis (1976, 1977) among others. When the function F in (6.1) depends on a parameter λ and a Hopf bifurcation for the corresponding kinetic equations (3.4) occurs at $\lambda = \lambda_o$, then under some circumstances, Cohen, Hoppensteadt, and Miura (1977) showed that solutions of (6.1) exist, which are represented by this Hopf solution modulated by a traveling wave envelop with large characteristic length. See also Miura (1977) and Ermentrout (1979).

8. SYSTEMS: SINGULAR PERTURBATION AND SCALING TECHNIQUES

8.1 Fast wave trains

Expanding target-like patterns are exhibited by the Belousov-Žabotinskiĭ reaction in laboratory situations. Far from the center, they are approximated by plane wave trains, so we can consider the existence of the latter type of solution to be experimentally verified.

One approach to the mathematical construction of trains has been through perturbation of spatially uniform but temporally oscillating solutions. This approach was successfully pursued by Kopell and Howard (1973), Ortoleva and Ross (1974), and Kopell (1977). Several existence proofs were given by Kopell and Howard; we shall be content with providing a formal perturbation analysis.

We begin by assuming that (6.1) has a periodic solution $u(x,t) = u_o(t)$ which depends on t only:

$$u_o' = f(u_o).$$

Let its period be T_o, and let $\sigma_o = 2\pi/T_o$. Then

$$u_o(t) = y_o(\sigma_o t), \tag{8.1}$$

where y_o is a 2π-periodic function of its argument.

We now look for solutions of (6.1) of the form

$$u(x,t) = y(\tau), \quad \text{where} \quad \tau = \sigma t - \alpha x, \tag{8.2}$$

and y is a 2π-periodic function of τ. All possible wave trains fit into this category, for if the wave length λ and period T are known, then the parameters α, σ, and the velocity $c = \lambda/T$ are determined by

$$\alpha = \frac{2\pi}{\lambda}, \quad \sigma = \frac{2\pi}{T}, \quad c = \frac{\sigma}{\alpha}.$$

Substituting (8.2) into (6.1), we obtain

$$\sigma y' = \alpha^2 D y'' + f(y). \tag{8.3}$$

Long wave-length trains are found, formally, by setting $\varepsilon = \alpha^2$,

$$y(\tau) = y_o(\tau) + \varepsilon y_1(\tau) + \ldots,$$

$$\sigma = \sigma_o + \varepsilon\sigma_1 + \ldots,$$

and taking ε to be small.

Our lowest order approximation is found by setting $\varepsilon = 0$; then

$$\sigma_o y_o' = f(y_o), \tag{8.4}$$

which is satisfied by the function y_o and parameter σ_o introduced in (8.1). We therefore use those as our first approximation.

To find the next approximation, we substitute the assumed expansion into (8.3) and look at terms of order ε. Alternatively, we may differentiate (8.2) with respect to ε, set $\varepsilon = 0$, and interpret

$$y_1 = \frac{\partial}{\partial\varepsilon} y\Big|_{\varepsilon=0}, \quad \sigma_1 = \frac{\partial\sigma}{\partial\varepsilon}\Big|_{\varepsilon=0}.$$

Either way, we obtain

$$\sigma_o y_1' + \sigma_1 y_o' = D y_o'' + f'(y_o) y_1,$$

or

$$L y_1 = D y_o'' - \sigma_1 y_o', \tag{8.5}$$

where $Ly \equiv \sigma_o y' - f'(y_o(\tau))y$.

This is to be solved under periodic boundary conditions $y_1(2\pi) = y_1(0)$. Now L, under these boundary conditions, is a Fredholm operator, and the equation

$$Ly = f,$$

for continuous periodic f, has a solution if and only if

$$(f,\psi) = \int_0^{2\pi} f(\tau)\psi(\tau)d\tau = 0$$

for every solution ψ of the homogeneous adjoint problem

$$L^*\psi = 0,$$

where

$$L^*\psi \equiv -\sigma_o\psi' - (f'(y_o(\tau)))^*\psi,$$

and $(f')^*$ denotes the transpose of f'.

Of course if L has no nullspace, then neither does L^*, and $y = L^{-1}f$ exists for all continuous f. But in the present case there is a nullspace, because $\phi \equiv y_o'$ satisfies $L\phi = 0$. This equation, in fact, is obtained by differentiating (8.4). Thus the nullspace of L has dimension at least 1. If the dimension is in fact greater than one, it will only be by accident. We now assume that the dimension is one. This will then also be the dimension of L^*'s nullspace, and there will be only one orthogonality condition $(f,\psi) = 0$ to be imposed. We further assume that 0 is an algebraically simple eigenvalue of L, so that $L^2y = 0$ has only the solutions $k\phi$. This implies

$$(\phi,\psi) \neq 0. \tag{8.6}$$

We now return to (8.5). The orthogonality condition now requires

$$(Dy_o'',\psi) - \sigma_1(\phi,\psi) = 0,$$

which in view of (8.6), is satisfied for precisely one value of σ_1.

This determines σ_1, and hence y_1 up to an additive multiple of ϕ:

$$y_1 = y_{10} + C\phi, \tag{8.7}$$

where C is an arbitrary constant.

It would be desirable to "pin down" y_1 by a normalization condition on y which would serve to specify C. The condition

$$(y,y_o') = 0 \tag{8.8}$$

serves this purpose. On the one hand, this is automatically satisfied for $\varepsilon = 0$, for then $y = y_o$, and

$$(y_o,y_o') = \frac{1}{2}\int_0^{2\pi} \frac{d}{d\tau}(y_o\cdot y_o)d\tau = 0.$$

And on the other hand, we may always subject the variable τ of our solution to a translation, still retaining the property of being a solution. So if $y(\tau,\varepsilon)$ is any family of solutions of (8.3) defined for small ε, differentiable in ε, and such that $y(\tau,0) = y_o(\tau)$, we may always replace it by a shifted family $y(\tau - s(\varepsilon),\varepsilon)$, where s is chosen so that (8.8) holds. This can be seen by applying the implicit function theorem to (8.8). It works because

$$\frac{\partial}{\partial s}\int_0^{2\pi} y(\tau - s,\varepsilon)\cdot y_o'(\tau)d\tau\Big|_{s=\varepsilon=0} = -\int_0^{2\pi} |y_o'(\tau)|^2 d\tau \neq 0.$$

So (8.8) brings no loss of generality. It implies the normalization conditions $(y_k,\phi) = 0$ for all the terms $y_k(\tau)$. In the case $k = 1$, this condition serves to determine the constant C in (8.7) uniquely.

Thus y_1 is determined uniquely. The same process may be continued to yield all higher order terms in y and σ.

The foregoing is a formal asymptotic expansion. There still remain two important questions:

(1) Does there exist an exact solution, approximated by the formal expansion?

(2) Is it stable?

The first question has been answered in the affirmative a couple of ways; see Kopell and Howard (1973) and Kopell (1977). The second is difficult. Although theoretically the question is amenable to Floquet analysis, the details are difficult and apparently have not been carried out except in the case of a special model system. The results for that special case can be reduced to a simple criterion for stability. The wave trains constructed by the above technique are sometimes, but not always, stable.

8.2 Sharp fronts (review)

It is commonplace, in many areas of applied analysis, to employ different space and time scales in different parts of the domain of functions being sought, because the latter exhibit particularly rapid or particularly slow variation at those places. The best known case is the phenomenon of a boundary layer for ordinary differential equations with a small parameter multiplying the highest order derivative term. But other cases abound, and in particular this technique is appropriate for analyzing a variety of types of wave fronts for RD equations, again involving one or more small parameters. The fronts so constructed have a "layer" of rapid variation, the abrupt part of the front, where scales should be used which are different from those in other parts of the front.

Probably the first use of singular perturbation techniques for propagating phenomena in RD systems and the like was in connection with the theory of combustion; see, for example, the review articles by Williams (1971) and Ludford (1976). Considerable work in this area is currently being done.

In biological applications, the construction of double-fronted traveling wave solutions for the FitzHugh-Nagumo equation of neurophysics by Casten, Cohen, and

Lagerstrom (1975) is notable (see also H. Cohen (1971)). The wave they obtained was a flat-topped pulse, representing a signal propagating along a nerve axon.

Carpenter (1977b, 1978) used similar concepts but quite different methods in her analysis of traveling waves of various kinds in excitable media such as a nerve axon.

In general two-component RD systems, the various possibilities for sharp fronts were categorized and partially analyzed in Fife (1976c), and many of them subjected to full asymptotic analysis in Fife (1977c). Kurland (1978) has given rigorous existence proofs for some structures of this type. Pismen (1978b) gave a different categorization of such fronts. Heuristic discussions of stability were given in that paper and in Fife (1976c). Other work using singular perturbation techniques or concepts includes that of Ortoleva and Ross (1976) and Feinn and Ortoleva (1977). The latter authors discuss many possible qualitative phenomena of this type for two- and higher-component systems.

Singular perturbation methods may be used to describe propagating fronts which are not standard traveling waves in the sense defined before, but rather have variable velocity, either because the domain is finite (Fife 1976c), or because the medium is inhomogeneous (Ball, Fife, and Peletier, in preparation).

8.3 Slowly varying waves (review)

Multiple scaling techniques have proved to be a fruitful approach to the description of wave-like solutions which are not strictly traveling waves, but which are locally approximated by plane wave trains. Howard and Kopell (1977, 1979) and Kopell (1977) studied such "slowly varying waves" in detail. In addition to the normal space and time variables x and t, they introduced slow space and time variables $X = \varepsilon x$ and $T = \varepsilon t$ for a small parameter ε. The solutions of (6.1) which they seek are of the form $u = U(\theta(x,t,\varepsilon),X,T,\varepsilon)$, U being 2π-periodic in θ. Defining $\Theta(X,T,\varepsilon) = \varepsilon\theta(x,t,\varepsilon)$, one assumes that Θ (as well as U) has an asymptotic expansion in powers of its last variable ε. This way, it is ensured that the lowest order (in ε) terms of θ are linear in x and t, which is requisite for local wave-train structure.

These authors apply this technique to study the problems of colliding wave trains (shocks), target patterns (circular waves emanating from a center), and boundary behavior (trains impinging on a boundary).

Multiple scaling approaches to some of these same problems were pursued by Ortoleva and Ross (1974).

Neu (1978) studied systems with small diffusion, whose kinetic equations have a time-periodic solution. With the diffusion present, the phase of the oscillator is adjusted by a function depending slowly on space and time, and which evolves according to a nonlinear parabolic equation.

8.4 Partitioning (review)

In Chapter 7, a discussion was given of the phenomenon of the appearance of small amplitude patterns through bifurcation from a uniform state. Perturbation techniques can be used to construct large amplitude patterns with the property that they divide up the reacting medium into subregions with sharply defined boundaries. The bounding partitions are defined as locations where one of the components of u undergoes an abrupt change, its variation in the interior of a subregion being relatively mild. The conditions for the appearance of such partitioned states, their evolution (through various stages) from initial data, and their stability, were studied in Fife (1976b) and Pismen (1978a) (part of Fife's argument was reproduced in Nicolis and Prigogine (1971), p. 207). A similar analysis for an infinite one-dimensional medium, showing periodic patterns, was given in Fife (1978b). Further possibilities were discussed in Fife (1977d). Rigorous existence results were given for analogous partitioning phenomena in one space variable in Fife (1976a), and in a two-dimensional inhomogeneous medium by Fife and Greenlee (1974).

This technique was applied to the problem of plankton patchiness by Mimura and Murray (1978) and Mimura, Nishiura, and Yamaguti (1978). Mimura (1978) is a related paper.

8.5 Transient asymptotics

An asymptotic solution of the initial value problem

$$u_t = \varepsilon u_{xx} + f(u),$$

$$u(x,0) = \phi(x),$$

was given by Berman (1978), using multiple scaling techniques. Results of this type were also obtained by Larson and Lange (1978), who illustrate the evolution of wave fronts from certain types of initial data ϕ.

9. REFERENCES TO OTHER TOPICS

9.1 Reaction-diffusion systems modeling nerve signal propagation

A pulse is a traveling structure whose profile approaches the same limit at $\pm\infty$. Thus, as distinguished from a front, the reacting medium returns to its original state after a pulse traverses it. Signals propagating along a nerve axon are very successfully modeled by pulse solutions of the Hodgkin-Huxley (HH) or FitzHugh-Nagumo (FHN) systems of reaction-diffusion equations, and this, in fact, is the context within which almost all the work on reaction-diffusion pulses has been performed. Excellent reviews of this work are available (Rinzel 1978a; Scott 1975, 1977; Hastings 1975; Troy 1978a; H. Cohen 1971), and so my comments here will be very brief.

Formally speaking, the HH system is a system of the type (6.1) with $n = 4$, and all elements of D zero except the upper left. The FHN system was devised as a simpler approximation to the HH system, and has $n = 2$, again with only one element of D nonzero. Finally, a third system, suggested by McKean (1970), has played an important role in modeling signal propagation. It is a modification of the FHN system, in that the nonlinear function appearing in the latter is replaced by a piecewise linear one. This was done with the idea that the resulting system should (1) be tractable analytically, and (2) have solutions with the same qualitative properties as the FHN system.

Both hopes have turned out to be justified for the McKean system; pulse solutions of it have been demonstrated, and their stability analyzed (Rinzel and Keller 1973, Rinzel 1975, Feroe 1978a). Moreover, the qualitative properties of the solutions, established analytically, agreed with those for the HH and FHN equations, which had been obtained by others numerically. The stability analysis in Rinzel and Keller (1973) was according the linearized criterion, and shows that some of the constructed pulses (the slow ones) are unstable. Feroe (1978a), building on the work of Evans (1972b, 1975), has recently established rigorously the stability of other pulses (the fast ones).

In addition, a family of wave trains was also constructed by Rinzel and Keller. Again by use of linear stability criteria, some members of the family were judged to be unstable, others probably stable.

The existence of pulse solutions for the HH and/or FHN systems, under various restrictions on the parameters of the equations, has been proved by Carpenter (1976, 1977b), Hastings (1976a,b), Conley (1975b,c), and Conley and Smoller (1976). As is the case for the McKean simplification, there exist two types of pulses: slow ones, thought to be unstable, and fast ones, thought to be stable. Establishing the stability properties of the pulses has proven to be a more difficult task even than their existence, and to date it cannot be said that a completely rigorous proof of stability of pulses for the HH and FHN equations has been given. Nevertheless, Evans (1972, 1975) has come a long way; he has reduced the problem to another criterion, which should be capable of verification, at least by use of a computer. At this point there is strong evidence in favor of stability for the fast pulse.

Carpenter (1976, 1977a,b), Conley (1975b,c), and Hastings (1974, 1976b) have also proved the existence of wave trains for the FHN equation, and Carpenter for the HH equation. In fact, Carpenter also obtained finite wave trains, or N-pulse solutions. See also Feroe (1978b).

Bell and Cook (1978) devised, on physiological grounds, an alternate system of Hodgkin-Huxley type with $n = 6$, and have proved the existence of pulses and wave trains using the methods of Carpenter.

Singular perturbation techniques have been applied to the FHN system when one of the parameters is small by Casten, Cohen, and Lagerstrom (1975). They construct pulses and wave trains of "traveling plateau" type, with abrupt transitions at the front and back of each pulse.

Mathematical work has been done on other questions associated with the nerve signal equations; see, for example, Schonbeck (1977), Rinzel (1977, 1978b), Evans and Shenk (1970), Evans (1972a, 1977), and Rauch and Smoller (1978).

9.2 Miscellaneous

A number of topics of high current interest in the theory of reaction-diffusion systems have not been mentioned in the preceding sections, or only briefly alluded to. As was indicated in the introduction, results within the field of chemical engineering were omitted because of the impossibility of adequate coverage here, and the existence of other surveys such as Aris (1975). The extensive work in Brussels on self-organizing phenomena has not been fully covered here, although many references were given. A more complete bibliography can be found in Nicolis and Prigogine (1977).

The analysis of spiral and target patterns in RD systems is also of high interest now, largely because of their appearance in laboratory demonstrations (Zaikin and Zhabotinsky 1970, Zhabotinsky and Zaikin 1973, Winfree 1972). Some of the relevant references are listed below. The dynamics of membranes and the mechanism of active transport across them is an exciting area of investigation. Again, some references are given below on this and other topics. The references are not complete; their ordering has no relation to their relative importance.

Target and spiral patterns in reacting and diffusing media

Winfree 1972, 1973a,b, 1978

Greenberg, Hassard, and Hastings 1978

Greenberg and Hastings 1978

Greenberg 1976, 1977

Ortoleva and Ross 1974

Ortoleva 1978 (the problem of coupling an oscillatory (perhaps chaotically) center to wave trains far away attacked by Padé approximants)

Cohen, Neu, and Rosales 1978 (rigorous existence of spiral waves for $\lambda - \omega$ systems under certain general conditions)

De Simone, Beil, and Scriven 1973

Howard and Kopell 1979

<u>Models for active transport across a membrane, front-like behavior, and other membrane phenomena</u>

Kernevez and Thomas 1973, 1975

Banks 1975

Caplan, Naparstek, and Zabusky 1973

Duban, Joly, Kernevez, and Thomas 1975

Brauner and Nicolaenko 1977

<u>Wave fronts for reation-diffusion systems, not covered in Section 8.2</u>

Zel'dovič 1948 (the first investigator of RD wave front models in combustion theory; later references not given here)

Conley 1978

Conley and Fife 1979 (population genetical context)

Tang and Fife 1979 (ecological equations of competing species)

Murray 1976 (model for the Belousov-Zhabotinskiǐ waves)

Troy 1978b (rigorous existence for Murray's waves)

Klaasen and Troy, in preparation (further results on Murray's model)

<u>Localized spatial structures</u>

Herschkowitz-Kauffman and Nicolis 1972

Boa and Cohen 1976

<u>Solitary waves, other than in neurophysiological contexts</u>

Field and Troy 1977; Troy 1977

Larson 1977a

<u>Other neurophysiological models exhibiting traveling waves and pattern formation</u>

Tuckwell and Miura 1978 (cortical waves)

Wilson and Cowan 1974; Cowan 1973 (integrodifferential models for large-scale neural activity)

Ermentrout and Cowan 1978a,b,c; Ermentrout 1979 (multi-faceted analyses of these models)

Integrodifferential equations in population genetics

Weinberger 1978, 1979

Integrodifferential equations, with emphasis on wave fronts, in epidemiology

Kendall 1965

Atkinson and Reuter 1976

Aronson 1976, 1977

Diekmann 1977a,b, 1978a,b

Mollison 1978

REFERENCES

H. Amann, 1977, Invariant sets and existence theorems for semilinear parabolic and elliptic systems, J. Math. Anal. Appl.

R. Aris, 1965, Prolegomena to the rational analysis of systems of chemical reactions, Arch. Rat. Mech. Analysis 19, 81-99.

R. Aris, 1975, The Mathematical Theory of Diffusion and Reaction in Permeable Catalysts I and II, Clarendon Press, Oxford.

D. G. Aronson, 1976, Topics in nonlinear diffusion, CBMS/NSF Lecture Notes, Regional Conference on Nonlinear Diffusion, Houston, 1976.

D. G. Aronson, 1977, The asymptotic speed of propagation of a simple epidemic, in: Nonlinear Diffusion, Res. Notes in Math. 14, Pitman Publishing Co., London, 1-23.

D. G. Aronson and H. F. Weinberger, 1975, Nonlinear diffusion in population genetics, combustion and nerve propagation, in: Proceedings of the Tulane Program in Partial Differential Equations and Related Topics, Lecture Notes in Mathematics 446, Springer, Berlin, 5-49.

D. G. Aronson and H. F. Weinberger, 1978, Multidimensional nonlinear diffusion arising in population genetics, Advances in Math., to appear.

C. Atkinson and G. E. H. Reuter, 1976, Deterministic epidemic waves, Math. Proc. Cambridge Phil. Soc. 80, 315-330.

J. F. G. Auchmuty, 1976, Positivity for elliptic and parabolic systems, FMRI Report #71, Univ. of Essex.

J. F. G. Auchmuty, 1978, Qualitative effects of diffusion in chemical systems, pp. 49-100 in Some Mathematical Questions in Biology, 9. Lectures on Math. in the Life Sciences, 10. (S. A. Levin, Editor) American Mathematical Society, Providence, Rhode Island.

J. F. G. Auchmuty and G. Nicolis, 1975, Bifurcation analysis of nonlinear reaction-diffusion equations, I. Evolution equations and the steady state solutions, Bull. Math. Biol. 37, 323-365.

J. F. G. Auchmuty and G. Nicolis, 1976, Bifurcation analysis of reaction-diffusion equations III. Chemical oscillations, Bull. Math. Biol. 38, 325-350.

J. F. G. Auchmuty and G. Nicolis, 1977, Time-periodic and wave-like solutions of reaction-diffusion equations, preprint.

I. Balslev and H. Degn, 1975, Spatial instability in simple reaction schemes, J. Theor. Biol. 49, 173-177.

H. T. Banks, 1975, Modeling and Control in the Biomedical Sciences, Lecture Notes in Biomathematics 6, Springer, Berlin.

C. Bardos and J. Smoller, 1978, Instabilité des solutions stationnaires pour des systemes de reaction-diffusion, C. R. Acad. Sci. Paris Ser. A-B, to appear.

G. I. Barenblatt and Ya. B. Zel'dovič, 1971, Intermediate asymptotics in mathematical physics, Uspehi Matem. Nauk SSSR 26, 115-129.

J. Bebernes and K. Schmitt, 1977, Invariant sets and the Hukuhara-Kneser property for systems of parabolic partial differential equations, Rocky Mountain J. of Math. 7, 557-567.

J. Bell and L. P. Cook, 1978, On the solutions of a nerve conduction equation, SIAM J. Appl Math. 35, 678-688.

A. Bensoussan, J. L. Lions, and G. Papanicolaou, 1978, Asymptotic Analysis for Periodic Structures, North-Holland Publishing Co., Amsterdam.

V. S. Berman, 1978, On the asymptotic solution of a nonstationary problem on the propagation of chemical reaction fronts, Doklady Akademiĭ Nauk SSSR 242, 265-267.

J. A. Boa, 1974, A Model Biochemical Reaction, Ph.D. Thesis, Caltech.

J. A. Boa, 1975, Multiple steady states in a model biochemical reaction, Studies in Appl. Math. 54, 9-15.

J. A. Boa and D. S. Cohen, 1976, Bifurcation of localized disturbances in a model biochemical reaction, SIAM J. Appl. Math. 30, 123-135.

J. R. Bowen, A. Acrivos, and A. K. Oppenheim, 1963, Singular perturbation refinement to quasi-steady state approximations in chemical kinetics, Chem. Eng. Sci. 18, 177-188.

M. D. Bramson, 1977, Maximal Displacement of Branching Brownian Motion, Ph.D. Thesis, Cornell University.

C. M. Brauner and B. Nicolaenko, 1977, Singular perturbation, multiple solutions, and hysteresis in a nonlinear problem, Singular Perturb. Bound. Layer Theory, Proc. Cong. Lyon 1976, Lecture Notes in Math. 594, Springer, Berlin, 50-76.

S. R. Caplan, A. Naparstek, and N. J. Zabusky, 1973, Chemical oscillations in a membrane, Nature 245, 364-366.

G. Carpenter, 1976, A mathematical analysis of excitable membrane phenomena, Proc. Third. Europ. Meeting on Cybernetics and Systems Research.

G. Carpenter, 1977a, Periodic solutions of nerve impulse equations, J. Math. Anal. Appl. 58, 152-173.

G. Carpenter, 1977b, A geometric approach to singular perturbation problems with applications to nerve impulse equations, J. Differential Equations 23, 335-367.

G. Carpenter, 1978, Bursting phenomena in excitable membranes, SIAM J. Appl. Math., to appear.

R. Casten, H. Cohen, and P. Lagerstrom, 1975, Perturbation analysis of an approximation to Hodgkin-Huxley theory, Quart. Appl. Math. 32, 365-402.

R. G. Casten and C. J. Holland, 1977, Stability properties of solutions to systems of reaction-diffusion equations, SIAM J. Appl. Math. 33, 353-364.

R. G. Casten and C. J. Holland, 1978, Instability results for reaction diffusion equations with Neumann boundary conditions, J. Differential Equations 27, 266-273.

N. Chafee, 1974, A stability analysis for a semilinear parabolic partial differential equation, J. Differential Equations 15, 522-540.

N. Chafee, 1976a, Asymptotic behavior for solutions of a one-dimensional parabolic equation with homogeneous Neumann boundary conditions, preprint.

N. Chafee, 1976b, Saddle point behavior and related phenomena for a one-dimensional parabolic equations, preprint.

N. Chafee and E. F. Infante, 1974, A bifurcation problem for a nonlinear partial differential equation of parabolic type, Applicable Anal. 4, 17-37.

J. Chandra and P. W. Davis, 1978, Comparison theorems for systems of reaction-diffusion equations, preprint.

K. N. Chueh, 1975, On the Asymptotic Behavior of Solutions of Semilinear Parabolic Partial Differential Equations, Ph.D. Thesis, Univ. of Wisconsin.

K. N. Chueh, C. Conley, and J. Smoller, 1977, Positively invariant regions for systems on nonlinear diffusion equations, Indiana Univ. Math. J. 26, 373-392.

D. S. Cohen, P. S. Hagan, and H. C. Simpson, 1978a, Traveling waves in single and multi-species predator-prey communities, SIAM J. Appl. Math., to appear.

D. S. Cohen, P. S. Hagan, and H. C. Simpson, 1978b, Spatial structures in predator-prey communities with hereditary effects and diffusion, preprint.

D. S. Cohen, F. C. Hoppensteadt, and R. M. Miura, 1977, Slowly modulated oscillations in nonlinear diffusion processes, SIAM J. Appl. Math. 33, 217-229.

D. S. Cohen, J. C. Neu, and R. R. Rosales, 1978, Rotating spiral wave solutions of reaction-diffusion equations, SIAM J. Appl. Math. 35, 536-547.

H. Cohen, 1971, Nonlinear diffusion problems, Studies in Appl. Math. 7, Math. Assoc. of America and Prentice Hall, 27-63.

C. Conley, 1975a, An application of Wazewski's method to a nonlinear boundary value problem which arises in population genetics, Univ. of Wisconsin Math. Research Center Tech. Summary Report 1444.

C. Conley, 1975b, On the existence of bounded progressive wave solutions of the Nagumo equation, preprint.

C. Conley, 1975c, On traveling wave solutions of nonlinear diffusion equations, Dynamic Systems Theory and Appl. (J. Moser, Editor), Lecture Notes in Physics 38, Springer-Verlag, Berlin and New York.

C. Conley, 1978, Isolated invariant sets and the Morse index, CBMS/NSF Ref. Conf. Ser. on Math. 38.

C. Conley and P. Fife, 1979, Critical manifolds, traveling waves and an example from population genetics, in preparation.

C. C. Conley and J. A. Smoller, 1976, Remarks on traveling wave solutions of non-linear diffusion equations, Structural Stability, The Theory of Catastrophes, and Applications in the Sciences (P. Hilton, Editor), Lecture Notes in Math. 525, Springer-Verlag, Berlin and New York.

E. Conway, D. Hoff, and J. Smoller, 1978, Large time behavior of solutions of systems of nonlinear reaction-diffusion equations, SIAM J. Appl. Math. 35, 1-16.

E. Conway and J. Smoller, 1977a, Diffusion and the predator-prey interaction, SIAM J. Appl. Math. 33, 673-686.

E. Conway and J. Smoller, 1977b, Diffusion and the classical ecological interactions: asymptotics, in: Nonlinear Diffusion, Proc. of NSF-CBMS Regional Conference on Nonlinear Diffusion, Research Notes in Math., Pitman, London.

E. Conway and J. Smoller, 1977c, A comparison technique for systems of reaction-diffusion equations, Comm. in Partial Differential Equations 2, 679-697.

J. Cowan, 1974, Mathematical models of large-scale nervous activity, Some Mathematical Questions in Biology, V, Lectures on Mathematics in the Life Sciences 6, Amer. Math. Soc., Providence, RI, 101-132.

M. Crandall and P. Rabinowitz, 1971, Bifurcation from simple eigenvalues, J. Functional Anal. 8, 321-340.

M. Crandall and P. Rabinowitz, 1973, Bifurcation, perturbation of simple eigenvalues, and linearized stability, Arch. Rat. Mech. Anal. 52, 161-120.

M. Crandall and P. Rabinowitz, 1977, The Hopf bifurcation theorem in infinite dimensions, Arch. Rat. Mech. Anal. 67, 53-72.

P. de Mottoni and F. Rothe, 1978, Convergence to homogeneous equilibrium state in generalized Volterra-Lotka systems with diffusion, preprint.

P. de Mottoni and A. Tesei, 1979, Asymptotic stability results for a system of quasilinear parabolic equations, Applicable Anal., to appear.

J. A. DeSimone, D. L. Beil, and L. E. Scriven, 1973, Ferroin-collodion membranes: dynamic concentration patterns in planar membranes, Science 180, 946-948.

O. Diekmann, 1977a, On a nonlinear integral equation arising in mathematical epidemiology, preprint.

O. Diekmann, 1977b, Limiting behaviour in an epidemic model, Nonlinear Anal., Theory, Methods, and Appl. 1, 459-470.

O. Diekmann, 1977c, Threshholds and travelling waves for the geographical spread of infection, Mathematical Centre Report TW 166, Amsterdam.

O. Diekmann, 1978, Run for Your Life. A Note on the Asymptotic Speed of Propagation of an Epidemic, Mathematisch Centrum, Amsterdam, preprint.

O. Diekmann and N. M. Temme, 1976, Nonlinear Diffusion Problems, Mathematisch Centrum, Amsterdam.

M. C. Duban, G. Joly, J. P. Kernevez, and D. Thomas, 1975, Hysteresis, oscillations, and morphogenesis in immobilized enzyme systems, preprint.

G. B. Ermentrout, 1979, in preparation.

G. B. Ermentrout and J. D. Cowan, 1978a, Large scale spatially organized activity in neural nets, SIAM J. Appl. Math., to appear.

G. B. Ermentrout and J. D. Cowan, 1978b, Temporal oscillations in neuronal nets, J. Math. Biol., to appear.

G. B. Ermentrout and J. D. Cowan, 1978c, Secondary bifurcation in neural nets, SIAM J. Appl. Math., to appear.

J. W. Evans, 1972a, Nerve axon equations: II. Stability at rest, Indiana Univ. Math. J. 22, 75-90.

J. W. Evans, 1972b, Nerve axon equations: III. Stability of the nerve impulse, Indiana Univ. Math. J. 22, 577-593.

J. W. Evans, 1975, Nerve axon equations: IV. The stable and the unstable impulse, Indiana Univ. Math. J. 24, 1169-1190.

J. W. Evans, 1977, Transition behavior at the slow and fast impulses, in: Nonlinear Diffusion, Research Notes in Mathematics 14, Pitman Publishing Co., London.

J. Evans and N. Shenk, 1970, Solutions to axon equations, Biophys. J. 10, 1090-1101.

M. Feinberg, 1972, Complex balancing in general kinetic systems, Arch. Rat. Mech. Anal. 49, 187.

M. Feinberg, 1978, Mathematical aspects of mass action kinetics, Chap. 1, R. H. Wilhelm Memorial Vol. on Chemical Reaction Theory.

M. Feinberg and F. J. M. Horn, 1974, Dynamics of open chemical systems and the algebraic structure of the underlying reaction network, Chem. Eng. Sci. 29, 775-787.

M. Feinberg and F. J. M. Horn, 1977, Chemical mechanism structure and the coincidence of the stoichiometric and kinetic subspaces, Arch. Rat. Mech. Anal. 66, 83-97.

D. Feinn and P. Ortoleva, 1977, Catastrophe and propagation in chemical reactions, J. Chem. Phys. 67, 2119.

W. Feller, 1936, Zur Theorie der stochastischen Prozesse, Math. Annalen 113, 113-160.

J. Feroe, 1978a, Temporal stability of solitary impulse solutions of a nerve equation, Biophys. J., to appear.

J. Feroe, 1978b, Existence and stability of multiple impulse solutions of a nerve equation, preprint.

R. J. Field and R. M. Noyes, 1974, Oscillations in chemical systems. IV. Limit cycle behavior in a model of a real chemical reaction, J. Chem. Phys. 60, 1877-1884.

R. J. Field and W. C. Troy, 1978, Solitary traveling wave solutions of the Field-Noyes model of the Belousov-Zhabotinskii reaction, Arch. Rat. Mech. Anal., to appear.

P. C. Fife, 1976a, Boundary and interior transition layer phenomena for pairs of second order differential equations, J. Math. Anal. and Appls. 54, 497-521.

P. C. Fife, 1976b, Pattern formation in reacting and diffusing systems, J. Chem. Phys. 64, 854-864.

P. C. Fife, 1976c, Singular perturbation and wave front techniques in reaction-diffusion problems, in: SIAM-AMS Proceedings, Symposium on Asymptotic Methods and Singular Perturbations, New York, 23-49.

P. C. Fife, 1977a, Stationary patterns for reaction-diffusion equations, in: Nonlinear Diffusion, Res. Notes in Math. 14, Pitman, London, 81-121.

P. C. Fife, 1977b, On modeling pattern formation by activator-inhibitor systems, J. Math. Biol. 4, 358-362.

P. Fife, 1977c, Asymptotic analysis of reaction-diffusion wave fronts, Rocky Mountain J. Math. 7, 389-415.

P. C. Fife, 1978a, The bistable nonlinear diffusion equation: basic theory and some applications, Proceedings, International Conference on Applied Nonlinear Analysis, Arlington, 1978, Academic Press, to appear.

P. C. Fife, 1978b, Asymptotic states for equations of reaction and diffusion, Bull. Amer. Math. Soc. 84, 693-726.

P. C. Fife, 1978c, Results and open questions in the asymptotic theory of reaction-diffusion equations, in: Nonlinear Evolution Equations, Academic Press (Proceedings of conference sponsored by Math. Res. Center, Univ. of Wisconsin, 1977), 125-139.

P. Fife, 1979, Long time behavior of solutions of bistable nonlinear diffusion equations, Arch. Rat. Mech. Anal., to appear.

P. C. Fife and W. M. Greenlee, 1974, Interior transition layers for elliptic boundary value problems with a small parameter, Usp. Matem. Nauk SSSR 24, 103-130; Russ. Math. Surveys 29, 103-131.

P. C. Fife and J. B. McLeod, 1977, The approach of solutions of nonlinear diffusion equations to travelling front solutions, Arch. Rat. Mech. Anal. 65, 335-361; Also: Bull. Amer. Math. Soc. 81 (1975), 1075-1078.

P. C. Fife and L. Peletier, 1977, Nonlinear diffusion in population genetics, Arch. Rat. Mech. Anal. 64, 93-109.

R. A. Fisher, 1930, The Genetical Theory of Natural Selection (Second Edition 1958), Dover, New York.

R. A. Fisher, 1937, The wave of advance of advantageous genes, Ann. of Eugenics 7, 355-369.

R. A. Fisher, 1950, Gene frequencies in a cline determined by selection and diffusion, Biometrics 6, 353-361.

W. H. Fleming, 1975, A selection-migration model in population genetics, J. Math. Biol. 2, 219-233.

A. Friedman, 1964, Partial Differential Equations of Parabolic Type, Prentice-Hall, Englewood Cliffs, NJ.

G. R. Gavalas, 1968, Nonlinear Differential Equations of Chemically Reacting Systems, Springer Tracts in Natural Phylosophy, Springer-Verlag, New York.

A. Gierer and H. Meinhardt, 1972, A theory of biological pattern formation, Kybernetika (Prague) 12, 30-39.

A. Gierer and H. Meinhardt, 1974, Biological Pattern Formation Involving Lateral Inhibition, Lectures on Math. in the Life Sciences 7, Amer. Math. Soc., Providence, RI, 163-183.

P. Glansdorff and I. Prigogine, 1971, Thermodynamics of Structure, Stability, and Fluctuations, Wiley-Interscience, New York.

J. I. Gmitro and L. E. Scriven, 1966, A physicochemical basis for pattern and rhythm, in: Intracellular Transport (K. B. Warren, Editor), Academic Press, New York and London.

J. M. Greenberg, 1973, A note on the Nagumo equation, Quart. J. Math. Oxford (2), 10.

J. M. Greenberg, 1976, Periodic solutions to reaction-diffusion equations, SIAM J. Appl. Math. 30, 199-205.

J. M. Greenberg, 1977, Axisymmetric time-periodic solutions to $\lambda - \omega$ systems, to appear.

J. M. Greenberg, B. D. Hassard, and S. P. Hastings, 1978, Pattern formation and periodic structures in systems modeled by reaction-diffusion equations, Bull. Amer. Math. Soc., to appear.

J. M. Greenberg and S. P. Hastings, 1978, Spatial patterns for discrete models of diffusion in excitable media, SIAM J. Appl. Math., to appear.

K. P. Hadeler, 1976, Travelling population fronts, in: Population Genetics and Ecology, Academic Press, New York.

K. P. Hadeler and F. Rothe, 1975, Travelling fronts in nonlinear diffusion equations, J. Math. Biol. 2, 251-263.

P. Hagan, 1979, Ph.D. Thesis, Caltech.

J. B. S. Haldane, 1948, The theory of a cline, J. Genetics 48, 277-284.

J. K. Hale, 1969, Dynamical systems and stability, J. Math. Anal. and Appl. 26, 39-59.

Alan Hastings, 1978, Global stability in Lotka-Volterra systems with diffusion, Journal of Mathematical Biology 6, 163-168.

S. P. Hastings, 1974, The existence of periodic solutions to Nagumo's equation, Quart. J. Math. Oxford, Ser. 25, 369-378.

S. P. Hastings, 1975, Some mathematical problems from neurobiology, Amer. Math. Monthly 82, 881-895.

S. P. Hastings, 1976a, On travelling wave solutions of the Hodgkin-Huxley equations, Arch. Rat. Mech. Anal. 60, 229-257.

S. P. Hastings, 1976b, On the existence of homoclinic and periodic orbits for the FitzHugh-Nagumo equations, Quart. J. Math. Oxford, Ser. 27, 123-134.

F. G. Heineken, H. M. Tsuchiya, and R. Aris, 1967, On the mathematical status of the pseudo-steady state hypothesis of biochemical kinetics, Math. Biosci. 1, 95-113.

D. Henry, 1976, Geometric Theory of Semilinear Parabolic Equations, Lecture Notes.

D. Henry, 1977, Gradient flows defined by parabolic equations, in: Nonlinear Diffusion, Research Notes in Mathematics 14, Pitman, London, 122-128.

M. Herschkowitz-Kaufmann, 1975, Bifurcation analysis of nonlinear reaction-diffusion equations II: Steady state solutions and comparison with numerical simulations, Bull. Math. Biol. 37, 589-635.

H. Herschkowitz-Kaufman and G. Nicolis, 1972, Localized spatial structures and nonlinear chemical waves in dissipative systems, J. Chem. Phys. 56, 1890-1895.

F. C. Hoppensteadt, 1966, Singular perturbations on the infinite interval, Trans. Amer. Math. Soc. 123, 521-535.

F. Hoppensteadt, 1975, Mathematical Theories of Populations: Demographics, Genetics, and Epidemics, Regional Conference Series in Applied Math. 20, Soc. for Industrial and Applied Mathematics, Philadelphia.

F. C. Hoppensteadt, 1976, A slow selection analysis of two locus, two allele traits, Theor. Population Biol. 9, 68-81.

F. Horn, 1972, Necessary and sufficient conditions for complex balancing in chemical kinetics, Arch. Rat. Mech. Anal. 49, 172.

F. Horn, 1973a, On a connexion between stability and graphs in chemical kinetics. I. Stability and the reaction diagram, Proc. R. Soc. Lond. A 334, 299.

F. Horn, 1973b, On a connexion between stability and graphs in chemical kinetics. II. Stability and the complex graph, Proc. R. Soc. Lond. A 334, 313.

L. N. Howard and N. Kopell, 1974, Wave trains, shock fronts, and transition layers in reaction-diffusion equations, SIAM-AMS Proceedings 8, Providence, RI, 1-12.

L. N. Howard and N. Kopell, 1977, Slowly varying waves and shock structures in reaction diffusion equations, Studies in Appl. Math. 56, 95-145.

L. N. Howard and N. Kopell, 1979, in preparation.

J. L. Ibañez and M. G. Velarde, 1977, Multiple steady states in a simple reaction-diffusion model with Michaelis-Menten (first order Hinshelwood-Langmuir) saturation law: The limit of large separation in the two diffusion constants, J. Math. Phys., to appear.

J. L. Ibañez and M. G. Velarde, 1978, Nonlinear reaction-diffusion on a sphere: a simple autocatalytic model with Michaelis-Menten (first order Hinshelwood-Langmuir) saturation law, J. Non-Equilib. Thermodyn. 3.

Y. Kametaka, 1976, On the nonlinear diffusion equation of Kolmogorov-Petrovskii-Piskunov type, Osaka J. Math. 13, 11-66.

Ya. I. Kanel', 1962, On the stabilization of solutions of the Cauchy problem for the equations arising in the theory of combustion, Mat. Sbornik 59, 245-288.

Ya. I. Kanel', 1964, On the stabilization of the solutions of the equations of combustion theory with initial data of compact support, Mat. Sbornik 65, 398-413.

J. P. Keener, 1976, Secondary bifurcation in nonlinear diffusion reaction equations, Studies in Appl. Math. 55, 187-211.

J. P. Keener, 1978, Activators and inhibitors in pattern formation, Studies in Appl. Math., to appear.

D. G. Kendall, 1965, Mathematical models of the spread of infections, in: Mathematics and Computer Science in Biology and Medicine, Medical Research Council, 218-225.

J. P. Kernevez and D. Thomas, 1973, Numerical analysis of immobilized enzyme systems, Rapport de Recherche No. 28, IRIA, Rocquencourt, France.

J. P. Kernevez and K. Thomas, 1975, Numerical analysis and control of some biochemical systems, Applied Math. and Optimization 1, 222-285.

H. Kielhöfer, 1976, On the Lyapunov-stability of stationary solutions of semilinear parabolic differential equations, J. Differential Equations 22, 193-208.

M. Kimura and J. F. Crow, 1969, Natural selection and gene substitution, Genet. Res. Camb. 13, 127-141.

A. Kolmogoroff, 1931, Über die analytischen Methoden in der Wahrscheinlichkeitsrechnung, Math. Annalen 104, 415-458.

A. N. Kolmogorov, I. G. Petrovskiĭ, and N. S. Piskunov, 1937, A study of the equation of diffusion with increase in the quantity of matter, and its application to a biological problem, Bjul. Moskovskovo Gos. Univ. 17, 1-72.

N. J. Kopell, 1977, Waves, shocks and target patterns in an oscillating chemical reagent, in: Nonlinear Diffusion, Research Notes in Mathematics 14, Pitman, London, 129-154.

N. Kopell and L. N. Howard, 1973, Plane wave solutions to reaction-diffusion equations, Studies in Appl. Math. 52, 291-328.

H. Kurland, 1978, Dissertation, Univ. of Wisconsin.

D. A. Larson, 1977a, On the existence and stability of bifurcated solitary wave solutions to nonlinear diffusion equations, J. Math. Anal. Appl.

D. A. Larson, 1977b, On models for two-dimensional spatially structured chemodiffusional signal propagation, preprint.

D. A. Larson, 1978, Transient bounds and time asymptotic behavior of solutions to nonlinear equations of Fisher type, SIAM J. Appl. Math. 34, 93-103.

D. A. Larson and C. G. Lange, 1977, Transient solutions to some weakly diffusive nonlinear diffusion equations, preprint.

R. Lefever, M. Herschkowitz-Kaufman, and J. W. Turner, 1977, Dissipative structures in a soluble nonlinear reaction-diffusion system, Physics Letters 60A, 389-391.

S. A. Levin, 1974, Dispersion and population interactions, Am. Nat. 108, 207-228.

S. A. Levin, 1976a, Spatial patterning and the structure of ecological communities, in: Some Mathematical Questions in Biology (S. A. Levin, Editor), VII, Lectures on Mathematics in the Life Sciences 8, Am. Math. Soc., Providence, 1-36.

S. A. Levin, 1976b, Population dynamic models in heterogeneous environments, Ann. Rev. Ecol. Syst. 7, 287-310.

S. A. Levin, 1978a, Population models and community structure in heterogeneous environments, pp. 439-476 in MAA Study in Mathematical Biology Vol. II: Populations and Communities (S. A. Levin, Ed.), Mathematical Association of America, Washington, D.C.

S. A. Levin, 1978b, Pattern formation in ecological communities, pp. 433-465 in Spatial Pattern in Plankton Communities (J.H. Steele, Ed.), Plenum, New York.

C. C. Lin and L. A. Segel, 1974, Deterministic Problems in the Natural Sciences, MacMillan Publishing Co., New York.

A. Lotka, 1920, Undamped oscillations derived from the law of mass action, J. Amer. Chem. Soc. 42, 1595.

G. S. S. Ludford, 1976, Combustion for large activation energy, Letters in Appl. and Eng. Sci. 4, 49-62.

M. Luskin and T. Nagylaki, 1979, Numerical analysis of weak random drift in a cline, preprint.

K. Maginu, 1975, Reaction-diffusion equations describing morphogenesis I: Waveform stability of stationary wave solutions in a one-dimensional model, Math. Biosci. 27, 17-98.

R. M. May, J. A. Endler, and R. E. McMurtrie, 1975, Gene frequency clines in the presence of selection opposed by gene flow, The Amer. Naturalist 109, 659-676.

H. P. McKean, 1970, Nagumo's equation, Advances in Math. 4, 209-223.

H. P. McKean, 1975, Application of Brownian motion to the equation of Kolmogorov-Petrovskii-Piskunov, Comm. Pure Appl. Math. 28, 323-331.

J. B. McLeod and P. C. Fife, 1979, A phase plane discussion of convergence to travelling fronts for nonlinear diffusion, to appear.

M. Mimura, 1978, Asymptotic behaviors of a parabolic system related to a planktonic prey and predator model, SIAM J. Appl. Math., to appear.

M. Mimura and J. D. Murray, 1978, On a planktonic prey-predator model which exhibits patchiness, preprint.

M. Mimura and Y. Nishiura, 1978, Spatial patterns for an interaction-diffusion equation in morphogenesis, J. Math. Biol., to appear.

M. Mimura, Y. Nishiura, and M. Yamaguti, 1978, Some diffusive prey and predator systems and their bifurcation problem, in: Proceedings, Int. Conf. on Bifurcation Theory and its Application to Scientific Disciplines, Ann. New York Acad. Sci., to appear.

R. M. Miura, 1977, A nonlinear WKB method and slowly-modulated oscillations in nonlinear diffusion processes, in: Nonlinear Diffusion, Research Notes in Mathematics 14, Pitman, London, 155-170.

H. J. K. Moet, 1978, A note on the asymptotic behavior of solutions of the KPP equation, SIAM J. on Math. Anal., to appear.

D. Mollison, 1977, Spatial contact models for ecological and epidemiological spread, J. Roy. Stat. Soc. B.

E. W. Montroll and B. J. West, 1973, Models of population growth, diffusion, competition and rearrangement, Synergetics, Cooperative Phenomena in Multi-Component Systems (H. Haken, Editor), 143-156.

J. D. Murray, 1975, Nonexistence of wave solutions for the class of reaction-diffusion equations given by the Volterra interacting population equations with diffusion, J. Theor. Biol. 52, 459-469.

J. D. Murray, 1976, On travelling wave solutions in a model for the Belousov-Zhabotinskii reaction, J. Theor. Biol. 56, 329-353.

J. D. Murray, 1977, Lectures on Nonlinear Differential-Equation Models in Biology, Clarendon Press, Oxford.

J. Nagumo, S. Arimoto, and S. Yoshizawa, 1962, An active pulse transmission line simulating nerve axon, Proc. Inst. Radio Engrs. 50, 2061-2070.

T. Nagylaki, 1975, Conditions for the existence of clines, Genetics 80, 595-615.

T. Nagylaki, 1976, Clines with variable migration, Genetics 83, 867-886.

T. Nagylaki, 1977, Selection in One- and Two-Locus Systems, Lecture Notes in Biomathematics 15, Springer, Berlin.

T. Nagylaki, 1978a, Clines with asymmetric migration, Genetics 88, 813-827.

T. Nagylaki, 1978b, Random genetic drift in a cline, Proc. Nat. Acad. Sci. 75, 423-426.

T. Nagylaki, 1978c. The geographical structure of populations, pp. 588-624 in MAA Study in Mathematical Biology, Vol. II: Populations and communities (S. A. Levin, Ed.), Mathematical Association of America, Washington, D.C.

T. Nagylaki, 1979, A diffusion model for geographically structured populations, J. Math. Biol., in press.

T. Nagylaki and J. F. Crow, 1974, Continuous selective models, Theor. Population Biol. 5, 257-283.

J. C. Neu, 1978, Chemical waves and the diffusive coupling of limit cycle oscillators, preprint.

A. C. Newell and J. A. Whitehead, 1969, Finite bandwidth, finite amplitude convection, J. Fluid Mech. 38, 279.

K. Nickel, The lemma of Max Müller-Nagumo-Westphal for strongly coupled systems of parabolic functional differential equations, MRC Tech. Report #1800.

G. Nicolis, 1974, Patterns of spatio-temporal organization in chemical and biochemical kinetics, SIAM-AMS Proc. 8, Providence, RI, 33-58.

G. Nicolis and I. Prigogine, 1977, Self-organization in Nonequilibrium Systems, Wiley-Interscience, New York.

P. Ortoleva, 1978, Dynamic Padé approximants in the theory of periodic and chaotic chemical center waves, J. Chem. Phys. 69, 300-307.

P. Ortoleva and J. Ross, 1974, On a variety of wave phenomena in chemical and biochemical oscillations, J. Chem. Phys. 60, 5090-5107.

P. Ortoleva and J. Ross, 1975, Theory of propagation of discontinuities in kinetic systems with multiple time scales: fronts, front multiplicity, and pulses, J. Chem. Phys. 63, 3398-3408.

H. G. Othmer, 1975, Nonlinear wave propagation in reacting systems, J. Math. Biol. 2, 133-163.

H. G. Othmer, 1977, Current problems in pattern formation, in: Lectures on Mathematics in the Life Sciences 9, Amer. Math. Soc., 57-85.

H. G. Othmer and L. E. Scriven, 1969, Interactions of reaction and diffusion in open systems, I & E. C. Fund. 8, 303-313.

L. A. Peletier, 1977, A nonlinear eigenvalue problem occurring in population genetics, in: Proceedings of Besançon Conference on Nonlinear Analysis, June, 1977.

L. M. Pismen, 1978a, Asymmetric steady states revisited, Chem. Eng. Sci., to appear.

L. M. Pismen, 1978b, Multiscale propagation phenomena in reaction-diffusion systems, J. Chem. Phys., submitted.

I. Prigogine and R. Lefever, 1968, Symmetry breaking instabilities in dissipative systems. II, J. Chem. Phys. 48, 1695-1700.

I. Prigogine and G. Nicolis, 1967, On symmetry-breaking instabilities in dissipative systems, J. Chem. Phys. 46, 3542-3550.

M. H. Protter and H. F. Weinberger, 1967, Maximum Principles in Differential Equations, Prentice-Hall, Englewood Cliffs, NJ.

J. Rauch and J. Smoller, 1978, Qualitative theory of the FitzHugh-Nagumo equations, Adv. in Math., to appear.

F. Riesz and B. Sz.-Nagy, 1955, Functional Analysis, Ungar Publishing Co., New York.

J. Rinzel, 1975, Neutrally stable traveling wave solutions of nerve conduction equations, J. Math. Biol. 2, 205-217.

J. Rinzel, 1977, Repetitive nerve impulse propagation: numerical results and methods, in: Nonlinear Diffusion, Research Notes in Mathematics 14, Pitman, London.

J. Rinzel, 1978a, Integration and propagation of neuroelectric signals, pp. 1-66 in MAA Study in Mathematical Biology, Vol. I: Cellular Behavior and the Development of Pattern (S. A. Levin, Ed.), Mathematical Association of America, Washington, D.C.

J. Rinzel, 1978b, Repetitive activity and Hopf bifurcation under point-stimulation for a simple FitzHugh-Nagumo nerve conduction model, J. Math. Biol. 5, 363-382.

J. Rinzel and J. B. Keller, 1973, Traveling wave solutions of a nerve conduction equation, Biophys. J. 13, 1313-1337.

F. Rothe, 1975, Über das asymptotische Verhalten der Lösungen einer nichtlinearen parabolischen Differentialgleichung aus der Populationsgenetik, Dissertation, Univ. of Tübingen.

F. Rothe, 1976, Convergence to the equilibrium state in the Volterra-Lotka diffusion equations, J. Math. Biol. 3, 319-324.

F. Rothe, 1978a, A simple system of reaction-diffusion equations describing morphogenesis. II. Existence and stability of equilibrium states, preprint.

F. Rothe, 1978b, Convergence to travelling fronts in semilinear parabolic equations, preprint.

F. Rothe and P. de Mottoni, 1978, A simple system of reaction-diffusion equations describing morphogenesis I: asymptotic behavior, Ann. Mat. Pura Appl., to appear.

D. H. Sattinger, 1971, Stability of bifurcation solutions by Leray-Schauder degree, Arch. Rat. Mech. Anal. 43, 154-166.

D. H. Sattinger, 1973, Topics in Stability and Bifurcation Theory, Lecture Notes in Mathematics 309, Springer, Berlin.

D. Sattinger, 1975a, A nonlinear parabolic system in the theory of combustion, Quart. Appl. Math., 47-61.

D. Sattinger, 1975b, Stability of traveling waves of nonlinear parabolic systems, in: Proc. of VIIth Inter. Conf. on Nonlinear Oscillations, East Berlin.

D. Sattinger, 1976, On the stability of waves of nonlinear parabolic systems, Advances in Math. 22, 312-355.

D. Sattinger, 1977, Weighted norms for the stability of traveling waves, J. Differential Equations 25, 130-144.

S. Sawyer, 1977a, Asymptotic properties of the equilibrium probability of identity in a geographically structured population, Adv. Appl. Prob. 9.

S. Sawyer, 1977b, Rates of consolidation in a selectively neutral migration model, Ann. of Probability 5, 486-493.

S. Sawyer, 1978a, Results for inhomogeneous selection-migration models, in preparation.

S. Sawyer, 1978b, A continuous migration model with stable demography, in preparation.

D. R. Schneider, N. R. Amundson, and R. Aris, 1972, On a mechanism for autocatalysis, Chem. Eng. Sci. 27, 895-905.

M. E. Schonbeck, 1977, Some results on the FitzHugh-Nagumo equations, in: Nonlinear Diffusion, Research Notes in Mathematics 14, Pitman, London, 213-217.

A. C. Scott, 1975, The electrophysics of a nerve fiber, Rev. Modern Phys. 47, 487-533.

A. C. Scott, 1977, Neurophysics, Wiley-Interscience, New York.

L. A. Segel and J. L. Jackson, 1972, Dissipative structure: an explanation and an ecological example, J. Theor. Biol. 37, 545-559.

L. A. Segel and S. A. Levin, 1976, Application of nonlinear stability theory to the study of the effects of diffusion on predator-prey interactions, Topics in Statistical Mechanics and Biophysics: A Memorial to J. L. Jackson, AIP Conference Proceedings, No. 27.

J. G. Skellam, 1951, Random dispersal in theoretical populations, Biometrika 38, 196-218.

M. Slatkin, 1973, Gene flow and selection in a cline, Genetics 75, 733-756.

A. N. Stokes, 1976, On two types of moving front in quasilinear diffusion, Math. Biosci. 31, 307-315.

Min Ming Tang and P. C. Fife, 1979, Propagating fronts for competing species equations with diffusion, Arch. Rat. Mech. Anal., to appear.

W. C. Troy, 1977a, A threshhold phenomenon in the Field-Noyes model of the Belousov-Zhabotinskii reaction, J. Math. Anal. and Appl. 58, 233-248.

W. C. Troy, 1977b, The disappearance of solitary travelling wave solutions of a model of the Belousov-Zhabotinsky reaction, Rocky Mountain J. of Math. 7, 467-478.

W. C. Troy, 1978a, Mathematical modeling of excitable media in neurobiology and chemistry, in: Periodicities in Chemistry and Biology, Adv. in Chem. Education, Academic Press.

W. C. Troy, 1978b, The existence of travelling wave front solutions of a model of the Belousov-Zhabotinskii chemical reaction, preprint.

H. C. Tuckwell and R. M. Miura, 1978, A mathematical model for spreading cortical depression, Biophysical J. 23, 257-276.

A. M. Turing, 1953, The chemical basis of morphogenesis, Phil. Trans. Roy. Soc. Lon. B237, 37-72.

J. J. Tyson, 1976, The Belousov-Zhabotinskii Reaction, Lecture Notes in Biomathematics 10, Springer, New York.

K. Uchiyama, 1977, The behavior of solutions of some non-linear diffusion equations for large time, preprint.

A. I. Vol'pert and S. I. Hudyaev, 1975, Analysis in Classes of Discontinuous Functions, and Equations of Mathematical Physics, Izd. Nauka, Moscow.

H. F. Weinberger, 1975, Invariant sets for weakly coupled parabolic and elliptic systems, Rend. Mat. 8 (VI), 295-310.

H. F. Weinberger, 1978, Asymptotic behavior of a model in population genetics, in: Nonlinear Partial Differential Equations and Applications, Lecture Notes in Mathematics 648, Springer, Berlin.

H. F. Weinberger, 1979, Asymptotic behavior of a class of discrete-time models in population genetics, Proc., Conference on Applied Nonlinear Analysis, Arlington, 1978, to appear.

F. A. Williams, 1971, Theory of combustion in laminar flows, Ann. Rev. Fluid Mech. 3, 171.

S. A. Williams and P-L. Chow, 1978, Nonlinear reaction-diffusion models for interacting populations, J. Math. Anal. and Appl. 62, 157-169.

H. R. Wilson and J. D. Cowan, 1973, A mathematical theory of the functional dynamics of cortical and thalamic nervous tissue, Kybernetika (Prague) 13, 55-80.

A. T. Winfree, 1972, Spiral waves of chemical activity, Science 175, 634-636.

A. T. Winfree, 1973a, Scroll-shaped waves of chemical activity in three dimensions, Science 181, 937-939.

A. T. Winfree, 1973b, Rotating solutions of reaction-diffusion equations in simply-connected media, SIAM-AMS Proceedings 8, Providence, RI, 13-31.

A. T. Winfree, 1978, Stably rotating patterns of reaction and diffusion, Theor. Chem. 4, Academic Press, New York, 1-51.

S. Wright, 1969, Evolution and the Genetics of Populations, Vol. II, Univ. of Chicago Press, Chicago.

A. N. Zaikin and A. M. Zhabotinsky, 1970, Concentration wave propagation in two-dimensional liquid-phase self-oscillating system, Nature 225, 535-537.

Ya. B. Zel'dovič, 1948, On the theory of propagating flames, Ž. Fiz. Him. 22, 27-48.

A. M. Zhabotinsky and A. N. Zaikin, 1973, Autowave processes in a distributed chemical system, J. Theor. Biol. 40, 45-61.

INDEX

Bio– mathematics

Managing Editors: K. Krickeberg, S. A. Levin

Volume 1:
Mathematical Topics in Population Genetics
Edited by K. Kojima
1970. 55 figures. IX, 400 pages
ISBN 3-540-05054-X

"...It is far and away the most solid product I have ever seen labelled biomathematics."
American Scientist

Volume 2: E. Batschelet
Introduction to Mathematics for Life Scientists
2nd edition. 1975. 227 figures. XV, 643 pages
ISBN 3-540-07293-4

"A sincere attempt to relate basic mathematics to the needs of the student of life sciences."
Mathematics Teacher

M. Iosifescu, P. Tăutu
Stochastic Processes and Applications in Biology and Medicine

Volume 3
Part 1: **Theory**
1973. 331 pages.
ISBN 3-540-06270-X

Volume 4
Part 2: **Models**
1973. 337 pages
ISBN 3-540-06271-8

Distributions Rights for the Socialist Countries:
Romlibri, Bucharest

"... the two-volume set, with its very extensive bibliography, is a survey of recent work as well as a textbook. It is highly recommended by the reviewer."
American Scientist

Volume 5: A. Jacquard
The Genetic Structure of Populations
Translated by B. Charlesworth, D. Charlesworth
1974. 92 figures. XVIII, 569 pages
ISBN 3-540-06329-3

"...should take its place as a major reference work.."
Science

Volume 6: D. Smith, N. Keyfitz
Mathematical Demography
Selected Papers
1977. 31 figures. XI, 515 pages
ISBN 3-540-07899-1

This collection of readings brings together the major historical contributions that form the base of current population mathematics tracing the development of the field from the early explorations of Graunt and Halley in the seventeenth century to Lotka and his successors in the twentieth. The volume includes 55 articles and excerpts with introductory histories and mathematical notes by the editors.

Volume 7: E. R. Lewis
Network Models in Population Biology
1977. 187 figures. XII, 402 pages
ISBN 3-540-08214-X

Directed toward biologists who are looking for an introduction to biologically motivated systems theory, this book provides a simple, heuristic approach to quantitative and theoretical population biology.

Springer-Verlag
Berlin
Heidelberg
New York